BASIC BIOLOGY
The Science of Living Things

BASIC BIOLOGY
The Science of Living Things

by Charles La Rue

Media Materials, Inc. • Baltimore, Maryland

THE AUTHOR

Charles J. La Rue holds a Ph.D. in science education, zoology, and botany from the University of Maryland, and an M.A. in zoology and botany from the University of Texas. He has spent fourteen years as a biology teacher, two years as a zoology and botany instructor at the college level, and fifteen years as a supervisor of elementary science in the Montgomery County (Maryland) Public Schools. He participated in the early development of BSCS Green Version of *Biological Science: An Ecological Approach*. He has co-authored fifty-five learning packages in science. He has also co-authored with Paul Brandewein *100 Investigations in Laboratory Science*. He serves as a consultant for the National Geographic Society for educational filmstrips, films, and books.

CONTRIBUTORS

Elissa D. Weinroth, Ph.D., made major contributions to the chapters on human health and on genetics. She served as consultant and preliminary editor for the text.

Alan Weinberg, M.Ed., a special education teacher, contributed to the chapters on the lives of plants and the systems of the human body.

Stephen McDaniel and Mary Angela McDaniel, former biology teachers, and now professional photographers, contributed to the chapters on human reproduction and on ecology.

Editor: Barbara Pokrinchak, Ed.D.
Editorial Consultant: M. E. Criste
Art: Rebecca Lyon Graphics Studio
Gregory Broadnax
Rimma Reyder

Printed in the U.S.A.   85   12   VG/ATB   5.0

ISBN: 0-86601-538-8

CONTENTS

Preface

What Is Biology?

Biology is a science that asks questions about what it means to be alive and about how living things do some of the things they do. Biology is about the relationships among all living things, and about their environments. Biology is about the insides of animals and plants—about how they are put together and how they work. Biology is about how animals and plants change in their own lifetime and how animals and plants have changed over long periods of time.

If students ask why they are going to study biology, they are off to a good start. Being curious is a common human trait. Humans have been asking and wondering about animals and plants for a long time. People in general have a natural curiosity about themselves and other living things.

The Purpose of This Book

Basic Biology: The Science of Living Things is a textbook designed for those students who need a direct, simplified presentation of the important concepts of biology. This book demonstrates the value of using the scientific system for classifying organisms. There is a name and a place for each organism. Readers will become more aware that the world is full of patterns. New information is discussed in such a way to show a connection with previous understandings. As the material unfolds, students move from the known to the unknown. They are encouraged to view their knowledge of basic life activities as being useful to them not only for the moment, but for tomorrow and the next day.

Organization of the Book

The main thread of the text is classification, animal biology, human biology, ecology, genetics, human reproduction, and evolution by natural selection. The fourteen chapters are each divided into meaningful key ideas. Each Key Idea section ends with questions to check understanding.

Chapter Conclusions

At the conclusion of each chapter there is a summary and a list of important words used in the chapter. The final exercise of the chapter is an extensive set of review questions. By answering these questions, students can determine if they have mastered the main ideas in the chapter.

Vocabulary

Concepts are presented in clear and simple language. Features that make the book easy to read are: large type, illustrations, generous headings, short sentences, and brief paragraphs. An effort was made to use familiar words and a comfortable style, while retaining essential biological terms.

Teacher's Guide

The *Teacher's Guide* that accompanies this text contains objectives, teaching suggestions, and complete answers. There is a substantial section of reproducible tests, laboratory activities, and reading exercises.

Manual of Laboratory and Language Activities

The accompanying student *Manual of Laboratory and Language Activities* provides additional opportunities for skill development in manipulation, observation, measurement, description, and classification. Reading and vocabulary exercises promote fuller comprehension of biological concepts.

Chapter 1

Describing How Living Things Are Alike

Chapter Goal:

To describe three ways in which living things are alike: in their cells, in their chemicals, and in their life activities.

Key Ideas:

- Living things are alike because they are all made up of basic units called cells.

- Living things are alike because they all have the same basic chemicals.

- Living things are alike because they all carry out the same kinds of basic life activities.

INTRODUCTION

Cells, the Basic Units

Living things are alike. Plants and animals are living things, and they are alike in several ways. They are alike in one way because they all have *CELLS*. Cells are the main parts of plants and animals. Cells are called *BASIC UNITS*. Cells, or basic units, are used for building body parts and for carrying out life activities.

Chemicals

Cells are living material that is made up of *CHEMICALS*. The same basic chemicals make up the bodies of plants and animals. Living things use the chemicals to stay alive. Living things need to have enough of the chemicals, and they need to have them in the right place at the right time. Water is a good example of a chemical that living things must have in order to stay alive.

Life Activities

Living things are also alike in how they stay alive. They all carry out similar *LIFE ACTIVITIES*. Getting food, eating, breathing, moving, digesting food, growing, reproducing, and sensing things, are among the chief life activities of animals and plants.

Living things are alike in three ways:
- They are all made up of basic units called cells.
- They are all made up of similar chemicals.
- They all carry out the same kinds of life activities.

Key Idea #1:

- Living things are alike because they are all made of basic units called cells.

CELLS: BASIC UNITS

What are cells? Cells are very small structures that are found only in living things. Cells are the smallest living whole things in life. They are invisible to the naked eye. They number in the millions and in the trillions. Cells are made of living matter. Cells are the living parts of plants and animals.

Size and Shape of Cells

Cells come in different sizes. They range from small to very small. They come in different shapes, such as triangles and squares. Cells also may be shaped like rectangles and circles. Cells may be long or short, or they may be wide or narrow.

Examples of Human Cells

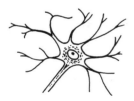 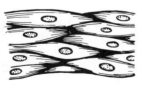 

Nerve cell   Muscle cells   Gland cells

Cell Functions

Cells carry out several functions. They play many parts in living things:

1. They cover (skin cells).
2. They help in movement (muscle cells).
3. They serve as linings (for the inside of blood vessels and body organs).
4. They send and receive signals (nerve cells).
5. They transport or carry (blood cells).

Cells in Plants and Animals

Cells are found in all parts of animals: in blood, bone, and muscle. Cells are in all parts of plants: roots, stems, leaves, and flowers.

Onion Skin Cells

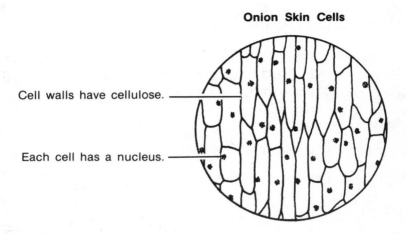

Cell walls have cellulose. ⸺

Each cell has a nucleus. ⸺

Cells Make Tissues

Cells are the basic living parts of plants and animals. Cells are used for building the different *TISSUES* of living things, such as muscle tissue, nerve tissue, and bone tissue. In animals there are muscle cells, blood cells, skin cells, nerve cells, and bone cells. Muscle cells are joined together in special ways to make different kinds of muscle tissue: leg muscles, arm muscles, stomach muscles, back muscles, hand and foot muscles, and heart muscles. Plants also have different kinds of cells, such as root cells, stem cells, and leaf cells. Inside a leaf there are different kinds of cells.

Groups of cells that are similar and act together to do some job are called *TISSUES*. Muscle cells work together to cause movement. The work is done by muscle tissue. Other tissues in living things are nerve tissue, bone tissue, and skin tissue. Living things are alike in the way they are put together with groups of cells called tissues.

Tissues Make Organs

Tissues join together to form organs. Tissues are the main parts of *ORGANS*. Organs are the main working parts of animals and plants. Your heart is an organ. Your lungs are organs. So are your stomach, liver, kidneys, eyes, ears, and bladder.

Living things are alike because they have cells that form tissues and because the tissues form organs. Organs are the main parts of living things.

Sense Organs   **Plant Organs**

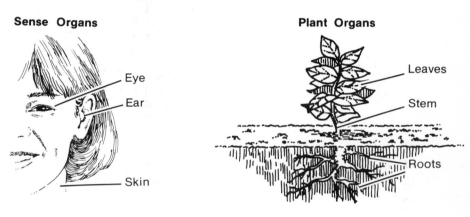

Eye
Ear
Skin

Leaves
Stem
Roots

CHECK YOUR UNDERSTANDING

1. Where can cells be found?
2. What are the shapes of cells?
3. What are three functions of cells?
4. What are three kinds of animal tissue?
5. What are tissues made of?
6. What are some of the kinds of plant cells that make tissue?
7. What are organs made of?
8. What are four kinds of organs?

> **Key Idea #2:**
> - Living things are alike because they all have the same basic chemicals.

BASIC LIFE CHEMICALS

Besides having cells, living things are alike because they have similar chemicals. Living things use these chemicals to stay alive.

Water, Water, Everywhere

Water is one of the chemicals that all living things use. Water is the most plentiful chemical in the bodies of animals and plants. You and other living things are mostly water. Water is found in each of the trillions of cells in the human body. For every one hundred pounds of your body, about sixty-eight pounds are water.

A Plant Cell

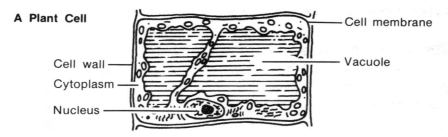

Cell membrane

Cell wall

Vacuole

Cytoplasm

Nucleus

Water — The Great Dissolver

Water is useful and it is necessary. *Life cannot exist without water.* Water *DISSOLVES* other chemicals. To dissolve means to change something from big pieces to smaller and smaller pieces. You can say that when something dissolves, it gets broken apart. Water breaks things apart. Cells are so small that the materials that go in and out of them must be dissolved. Water is the chemical that does the dissolving. When material is dissolved into small and invisible pieces, it can move more easily from one cell to another cell.

Chemicals of Life — 7

Salt Water in Cells

The water in your cells is not pure or fresh water. It is *SALT WATER*. It is a weak solution of salt water. Regular table salt and some other salts are the chemicals dissolved in the water of your body. The water in all animals is a mix of salts and water. Even fish found in fresh water streams and ponds have water with salt in their bodies.

Carbohydrates, Fats, and Proteins

Other important chemicals found in animals are *CARBOHYDRATES, FATS,* and *PROTEINS*. Each of these common chemicals does some job in the body of a living thing. *Carbohydrates* contain carbon, hydrogen, and oxygen. Look again at the word *carbohydrates*. The first part, *carbo-*, stands for carbon, and the second part, *hydrates*, stands for hydrogen and oxygen. Carbohydrates are chemicals that are used by you and all other living things as a *source of energy*.

Chemicals for Energy

Energy is the stuff that makes things go. Energy comes from fuel. Carbohydrates are "fuel chemicals." Carbohydrates in your body work like gasoline in your car. Gasoline from the fuel tank in your car gets up to the engine, where it is broken apart and energy is released. This released energy runs the engine. When the carbohydrates break apart in your body, energy is released. This energy is what makes you run. The same thing happens in other animals. Plants use energy, too.

Fats for Fuel

FATS are important chemicals found in foods such as beef, butter, cheese, and peanut butter. Fats provide the body with a source of energy. Fats supply a source energy because they are fuel chemicals.

Proteins for Life

PROTEINS are another kind of chemical. Proteins are part of all living things. Proteins do two main things:

1) They help to build and repair body parts.
2) They help control body activities such as heart rate, and the breaking down of food in the stomach and small intestine.

Nutrients and Your Diet

Keeping your body going is not a simple job. You must get a regular supply of carbohydrates, proteins, and fats from your food. It is important to get a *variety* of food in your diet. Each kind of food that you eat provides different chemicals — chemicals for energy and chemicals for growth. In addition to carbohydrates, proteins, and fats, your body also needs minerals, vitamins, and water. These chemicals — carbohydrates, proteins, fats, minerals, vitamins, and water — are all called *NUTRIENTS*. Living things need a fresh supply of these nutrients every day.

Some Sources of Nutrients

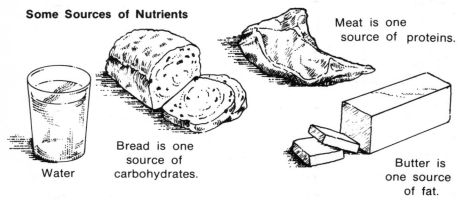

Meat is one source of proteins.

Water

Bread is one source of carbohydrates.

Butter is one source of fat.

Living Things Are Alike

Can you see how living things are alike? Living things are made up of cells. Living things are made up of the same chemicals. Living things need and use the same chemicals. Living things need nutrients.

CHECK YOUR UNDERSTANDING

1. Why do living things need water?
2. What kind of water is in your cells?
3. What are three important chemicals in animal bodies? Unscramble these words:
   a) DHCRASEABROTY   b) ETSPNRIO   c) TSFA
4. Of what special use are carbohydrates in your body?
5. What is the job of proteins in your body?
6. What are six nutrients that are chemicals of life?

Key Idea #3:

- Living things are alike because they all carry out the same kinds of basic life activities.

BASIC LIFE ACTIVITIES

All living things carry on the same kinds of activities. Getting food, sensing things, getting oxygen, digesting food, moving, growing, and reproducing are some of the basic *LIFE ACTIVITIES* of animals and plants. If all living things are made of cells, and if they all have similar sets of chemicals, then it should not be a surprise to you that all living things carry out the same kind of life activities.

Getting Food

A common example of a life activity is *GETTING FOOD*. Animals must get their food by eating plants or by capturing other animals. Bees and elephants, humans and whales, birds and sharks, all must get their food. Plants make their own food. Roses and maple trees, grass and corn, grapevines and orange trees, all make their own food.

Some animals eat other animals for food.

Green plants can make their own food.

Movement

MOVEMENT is another life activity that is common to all plants and animals. Plants do not move from one place to another, but they still move. Animal movement is easy to see. Plants have roots that hold them in one place, but their parts bend or move toward light. Besides outward movement, there is constant movement inside of living things. The insides of plants and animals are always changing. Movement goes on and on. Blood is circulating, food is digesting, and breathing is taking place. Movement is one of the basic life activities.

Growing

GROWING is one of the most obvious basic life activities. You have grown. You may have been a small or a large baby, but you have now grown to a larger person. You are still growing. You will continue to grow until you reach your adult size. Most living things go through a similar pattern, small to large, stopping at the adult state. Growing is part of being alive. It is one of the basic life activities.

Sensing and Reacting

SENSING THINGS and *REACTING* are basic life activities. All living things behave in certain ways. Animals and plants have tissues and organs that pick up signals from their surroundings. They have special parts that act as sensors. Plants and animals react or respond to the signals. The signals may be light, sound, smells, or touch. For example, moths come to lights at night; fish come to the top of a fish tank for food; and your dog comes to you at the sound of your voice. Plants turn toward the light. Their roots grow downward to the force of gravity. Flowers open in the morning light and close as the sun goes down. Sensing and reacting are things that all living things do. Sensing and reacting are basic life activities.

A sunflower turns toward the sun.

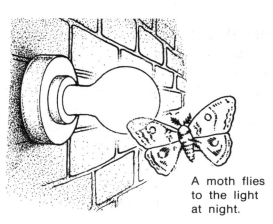

A moth flies to the light at night.

Using Food and Removing Wastes

DIGESTION is a life activity that changes the food you eat into a useful form for cells to use. *RESPIRATION* is another basic life activity. It causes oxygen to meet with the chemicals in food to release energy for body cells to do work. Respiration also gets rid of some cell waste matter. *EXCRETION* is the basic life activity that involves getting rid of solid and liquid waste matter from the body.

Development

DEVELOPMENT is another one of the basic life activities. For example, a puppy goes through changes that are a part of growing up. You know that what was at one time a small ball of fur develops a different shape snout, a very special tail, and an entirely new voice box (from a whiner to a barker). Development means getting bigger, but it also means getting different. Tadpoles develop by stages into frogs and toads. They get bigger and at the same time get different parts. So development is another thing that all living things go through — it is another one of the basic life activities.

Reproduction

REPRODUCTION is a basic life activity. Reproduction is not always something that individuals can do alone. Some lower plants and animals just grow to a certain size and then split. Most reproduction, however, involves two parents. All living things have reproduction as one of their basic life functions.

CHECK YOUR UNDERSTANDING

1. What is the difference in the way that animals and plants get food?
2. What is the main difference between the way animals and plants move?
3. For how long does a person continue to grow?
4. What is meant by "sensing things and reacting"?
5. What are three life activities that have to do with using food and removing wastes?
6. What changes does a puppy go through in the basic life activity called "development"?

SUMMARY OF CHAPTER 1

Living things are alike in three ways. They have basic units called cells; they have sets of similar chemicals; and they carry out the same kinds of basic life activities.

Cells are counted in the millions. They come in many sizes and shapes. Cells do a great deal of work in the body. They cover, help movement, line organs, send signals, and transport materials. Cells are the building blocks of tissues and organs.

The chemicals of life are water, carbohydrates, fats, and proteins. These chemicals serve as energy, and as building and repairing materials.

All plants and animals carry out the same basic life activities. These include getting food, moving, growing, sensing, reacting, using food, removing waste, developing, and reproducing.

WORDS TO USE

1. cells
2. cell functions
3. tissues
4. basic units
5. organs
6. basic chemicals
7. proteins
8. food
9. water
10. carbohydrates
11. fuel chemicals
12. dissolves
13. salt water
14. fats
15. nutrients
16. growing
17. sensing
18. respiration
19. reproducing
20. movement
21. reacting
22. excretion
23. basic life activities
24. digestion
25. developing

• REVIEW QUESTIONS FOR CHAPTER 1 •

1. What are three ways in which living things are alike?

2. What are three things that you can tell about cells?

3. What are the main functions that cells carry out?

4. What are the names of three kinds of animal tissue?

5. What kinds of cells are found in plants?

6. What is the definition of the term "tissue"?

7. How are organs related to tissues?

8. Tell three things about the chemical WATER as it relates to living animals.

9. What is the special use for carbohydrates in living things?

10. What are some foods that provide fats in the body?

11. What is it that fats do for animals?

12. What are the main things that proteins do for animals?

13. What are the basic life activities?

14. How is movement different in plants and in animals?

15. How is "development" different from just plain growing?

Chapter 2

Organizing All Natural Things

Chapter Goals:

To explain the basic differences between living and non-living things.

To explain why the living world is divided into four separate groups.

Key Ideas:

- Living things differ from non-living things in their characteristics and properties.

- The living world is divided into four major groups: plants, animals, protists, and fungi.

INTRODUCTION

Living and Non-Living

What is biology? *BIOLOGY* is the study of living things or organisms. What is an *ORGANISM?* An organism is a living thing that must breathe, move, grow, react, reproduce, digest, and get food. An organism must carry out the basic life activities. These basic life activities make living things more complex than non-living things.

The non-living world is made up of things that are solids, liquids, and gases. The mountains, rivers, oceans, and the air all belong to the non-living world.

Plants, Animals, Protists, and Fungi

Scientists have classified all the organisms in the world into four large groups. These are plants, animals, protists, and fungi. Most of the living things that you know are either plant or animal. But, there are two groups that you probably do not know as well. These are the protists and fungi. Protists and fungi are separate because they are hard to fit into either a plant or an animal group. But they do belong to the living world, because they carry out the basic life activities.

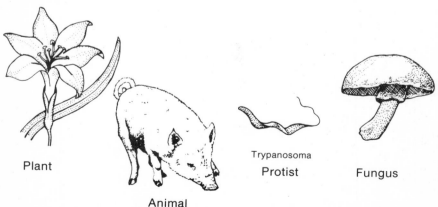

Plant

Animal

Trypanosoma
Protist

Fungus

> **Key Idea #1:**
>
> • Living things differ from non-living things in their characteristics and properties.

LIVING OR NON-LIVING?

The entire world of *NATURAL THINGS* may be taken apart and put into just two groups — *LIVING THINGS* into one group and *NON-LIVING THINGS* into the second group. Things are either plant, animal, protist, or fungi; or they are rocks, soil, water, or gases like oxygen and carbon dioxide.

Non-Living Things

It is very simple for a thing to qualify as non-living. A non-living thing does not carry out any of the basic life activities. The non-living things in the world are all of those basic materials that we call *solids, liquids,* and *gases.* Sand, water, and air qualify as non-living. So do rocks, minerals, and vitamins. None of these things have any of the basic life activities.

Properties

In order to describe non-living things, we talk about their *PROPERTIES*. Properties of non-living things are what you can see or find out about an object. The color of a rock is one property. How a rock breaks into pieces is another property. The degree of hardness of a rock is still another property. Each rock or mineral has a different degree of hardness. Each kind will scratch some things and not others because of its hardness. A scale of hardness can be made to identify each kind of rock or mineral. Did you know that a diamond is the hardest mineral of all? It is used in industry because of that property.

Living Things

To qualify as a living thing there must be *BASIC LIFE ACTIVITIES* such as getting food, moving, and growing. Plants and animals carry out the basic life activities. Plants and animals make up that part of the world we call the living world. The living world is made up of *ORGANISMS,* and plants and animals are organisms.

The first part of the word *organism* is *ORGAN.* Plants and animals have organs, or special parts, that are used to carry out the basic life activities. Non-living things do not have organs, and they are not qualified to be organisms.

Basic Differences among Living Things

Another basic difference between living and non-living things is that living things are more complex or complicated than non-living things. Also, there are more differences between *different* living things.

No two people are alike, not even identical twins. All of the puppies in a new litter are not alike. A characteristic of living things is that each one is different.

Think about some living things that seem to be very much alike. In what ways are these things alike? In what ways are they different? How easy or difficult was it to find the differences?

Now think of some non-living things in the *natural world* that seem to be nearly identical. Pieces of coal, sand grains, and snowflakes may have come to your mind. There are fewer differences between grains of sand and snowflakes than between the puppies in a litter.

The idea that "no two things are exactly alike" is easier to see when you think about living things. The *basic life activities* of living things make living things more complex or complicated than non-living things. Finding differences between living and non-living things is one way to divide the whole world into two parts.

CHECK YOUR UNDERSTANDING

1. Into what two groups can everything in the world be divided?
2. How does something in the world qualify as a living thing?
3. What is an organism?
4. What are two differences between living and non-living things?

Key Idea #2:

- The living world is divided into four major groups: plants, animals, protists, and fungi.

FOUR MAJOR KINGDOMS

At one time, scientists divided the world of living things into two main groups or *KINGDOMS,* the plant kingdom and the animal kingdom. Later, when biologists began to use the microscope, they found tiny organisms that did not fit into either the plant or animal kingdom. A new kingdom was formed in order to have a place for these *MICROORGANISMS.* This kingdom was called *PROTISTS.* Later, another kingdom called *FUNGI* was added. In this system of classification, there are four main kingdoms in the living world.

The Plant and Animal Kingdoms

The two groups with which you are most familiar are the *PLANTS* and *ANIMALS.* What are some ways to tell the difference between an animal and a plant?

Animals and plants are quite different.

The squirrel and the dandelion are both living things, but they are different. The squirrel can move about. The dandelion stays in one place. The squirrel has to find its food. The dandelion makes its own food.

Getting Food

Animals and plants differ in the way they get their food. Animals get their food from other living things. They eat plants, or they eat other animals. Animals move around to capture or gather their food. Tigers chase and capture antelopes, bears catch fish, and hawks capture field mice.

Plants make their own food from the materials they take in through their roots. They also use the energy they get from the sun. Trees, shrubs, grasses, and vines are easy to recognize as plants. They stay in one place and make their own food.

Protists and Fungi

The world of living things has two other groups of organisms that you probably do not know as well as plants and animals. These are the protists and fungi. They are in separate kingdoms because they are hard to fit into either a plant or animal group. None of these organisms resembles the things that you call plants and animals. But they do belong to the living world, because they carry out the basic life activities.

What Protists Are Like

Most *PROTISTS* are *MICROORGANISMS*. That is, they are very small and can only be seen under the microscope. The most common of the protists are the *BACTERIA* and the *PROTOZOANS*. Protozoans include parameciums, euglenas, and amebas. Some protists have both plant and animal traits. They act like animals by moving about, but they make their own food like the plant kingdom.

Bacteria   Paramecium   Euglena   Ameba

The Importance of Protists

Although protists are very small, there are very many of them, and they are important to humans. Protists include bacteria that cause diseases and bacteria that are useful to people. Cheese and yogurt are made because of the action of bacteria. Bacteria serve as food for other small organisms, which in turn are eaten by larger animals. Finally, some of these animals end up in the *food chains* that include humans. For example, many crustaceans or shellfish eat materials worked upon by bacteria. Shellfish, in turn, may become food for humans.

The Kingdom of Fungi

The fourth kingdom in the four kingdom classification system is the kingdom of *FUNGI*. (*Fungi* is the plural of *fungus*.) At one time fungi were classified with plants. However, they do not carry out one of the basic life activities of plants: they do not make their own food. Instead they take in, or *ABSORB,* food from other organisms. Neither do fungi carry out one of the basic life activities of animals and protists. They do not move around by themselves; they stay in one place.

Many fungi are *PARASITES.* Parasites are organisms that live on other organisms and depend upon them for their food. Some common fungi are mushrooms and molds.

The Great Decomposers

The most important job of the fungi is to *DECOMPOSE* or break down materials around them, so that they can be used by other organisms. Fungi decompose waste matter such as dead animals and plants, change them into usable chemicals, and return these chemicals to the soil, where they can be used by other organisms.

CHECK YOUR UNDERSTANDING

1. What are the four main kingdoms of the living world?
2. What are two differences between plants and animals?
3. Why do scientists put protists and fungi into separate kingdoms?
4. What are microorganisms?
5. How are protists similar to both plants and animals?
6. How do fungi get their food?
7. How are fungi helpful to other organisms?

SUMMARY OF CHAPTER 2

The two main categories of objects in the world are living things and non-living things. Living things, or organisms, can move, breathe, get food, and perform the basic life activities; while non-living things are unable to do so. Living things are also more different from each other than non-living things are.

There are billions of organisms in the world. They are easier to understand if we separate them into categories. Organisms are divided into plants, animals, protists, and fungi.

WORDS TO USE

1. biology	7. properties	14. microscopic
2. organism	8. organ	15. bacteria
3. natural things	9. kingdoms	16. protozoans
4. living things	10. plants	17. fungi
5. non-living things	11. animals	18. absorb
6. basic life activities	12. microorganism	19. parasites
	13. protists	20. decompose

• REVIEW QUESTIONS FOR CHAPTER 2 •

1. What are two main categories in the world of natural things?

2. What three basic materials are non-living things made of?

3. What are the properties of this book?

4. Besides basic life activities, what else makes living things different from non-living things?

5. What are the four major groups of organisms in the world?

6. What are some characteristics of each of these four groups?

7. How can bacteria be helpful to humans? Name three ways.

8. How can bacteria be harmful to humans?

9. What is a parasite?

10. What is the most important job of the fungi?

Chapter 3

Organizing Animals

Chapter Goals:

To understand how a system for organizing animals works.

To make use of the system of classification by organizing some familiar animals.

Key Ideas:

- Animals are classified into different groups, according to their similar structures.

- Each animal has a special, two-word name.

- A large part of the animal kingdom is made up of animals with backbones. They are called vertebrates.

- Animals without backbones make up a major group in the animal kingdom. They are called invertebrates.

INTRODUCTION

There is a system for classifying all animals by names and categories. An example of how the system works can be seen in the case of the shoebill stork. The shoebill stork is a large wading bird. This very large bird is a relative of other storks and herons. All of the wading birds in the world have *SIMILAR STRUCTURES* that are useful to their life in and around water. But within the whole group, each one of the wading birds has structures that separate it from other kinds of wading birds. The shoebill is so different from its relatives that it is called "one of a kind" by the scientists who study birds. The shoebill has a scientific name that identifies it. The shoebill is called *Balaeniceps rex*. It is the only bird in the world with that name.

Shoebill Stork — A wading bird

Balaeniceps rex

In the system for organizing animals there is a special name for each kind of animal. This chapter is about that naming system. You will see why the system is useful and important for everyone.

> **Key Idea #1:**
>
> - Animals are classified into different groups, according to their similar structures.

USING SIMILAR STRUCTURES TO CLASSIFY

Birds of a Feather

It is easy to classify birds as a single group of animals because birds have feathers. Feathers are structures that belong to birds and to birds alone. No other animals have feathers. Feathers are used for flying. Feathers also keep in the body heat of birds. The single structure, feathers, separates the birds of the entire world into one group. To this group belong robins, sparrows, mockingbirds, bluejays, crows, starlings, ducks, and brown-headed cowbirds. The list could go on and on.

Animals with Fur

Another large group of animals are those with fur. Fur belongs to animals like bears, raccoons, squirrels, deer, lions, tigers, rabbits, rats, cats, dogs, elephants, orangutans, and giraffes. Fur helps insulate the body and keep in body heat. In the system of classification, animals that have fur and feed their babies milk are called *MAMMALS*. Some of the common mammals are dogs, cats, squirrels, whales, bears, and *HUMANS*.

Animals with Fur

Raccoon
Procyon lotor

Lion
Felis leo

Scales

Fish have scales. Scales form the body covering of fish. If you have scaled a fish, you know about the hundreds and hundreds of small scales that are on the outside of the fish's body. Scales belong to fish — not to birds and not to mammals. Bass, trout, sunfish, perch, pickerel, tuna, salmon, goldfish, carp, and guppies: every one of these is a fish, and every one has scales.

But just one moment, please. Snakes, lizards, and turtles have scales, too! So, having scales as a body structure does not mean *"for fish only!"* Snakes, lizards, and turtles are *REPTILES*. How can we separate the reptiles from the fish? We need to find some separate structures for each kind of organism.

Scientists have found that fish have hearts that have *ONE CHAMBER*. Also, fish must live in water. Reptiles have hearts with *TWO CHAMBERS*. They may live on land as well as in the water. In addition, reptiles lay eggs with soft shells. Scales alone are not useful in separating fish from reptiles. So two large groups of animals, fish and reptiles, must be separated in other ways.

CHECK YOUR UNDERSTANDING

1. What body structure do birds have in common?

2. What other body structures are used to classify animals into groups?

3. What animals have scales?

4. What do scientists use to separate fish and reptiles into two different groups?

> **Key Idea #2:**
>
> • Each animal has a special, two-word name.

ALL ANIMALS FIT SOMEWHERE IN THE SYSTEM

Each kind of animal has a place of its own in the system of classification. The system keeps track of thousands and thousands of animals. The system works on a simple plan of *levels*. The highest level is the level of *KINGDOM*. Once an organism is placed in the *ANIMAL KINGDOM*, then the scientist needs to decide what kind of animal it is.

The scientist asks questions about the structure of the animal. Finally, the information is narrowed down to the two lowest levels. Then the animal receives its scientific name of two words. These names are in the Latin language.

Scientific Names

Homo sapiens

In the case of the human animal, the scientific name is *Homo sapiens.* The first word, *Homo,* tells what genus a human is. The second word, *sapiens,* tells to what species a human belongs. Whenever you see the word *Homo sapiens,* you can be sure that this is another name for "human."

The common pet that you call "dog" is in the animal kingdom. Its scientific name is *Canis familiaris.* It is in the genus *Canis.* Its species name is *familiaris.*

Canis familiaris

The little house mouse is known to scientists as *Mus musculus*. It is in the genus *Mus*. Its species name is *musculus*. There is no other animal in the world by that name. If you find another animal that is very much like a house mouse, but is not really a house mouse, you will have to give it another name.

House mouse
Mus musculus

The Seven Levels of Classification

The chart below shows the seven levels that scientists use to classify each animal. These levels are the main categories or groups. Notice that the highest level is *KINGDOM*. The lowest levels, 6 and 7, are *GENUS* and *SPECIES*.

LEVELS OF CLASSIFICATION USED WITH THREE ANIMALS

LEVELS	HUMANS	DOGS	HOUSE MOUSE
1. KINGDOM	Animal	Animal	Animal
2. PHYLUM	Chordate	Chordate	Chordate
3. CLASS	Mammals	Mammals	Mammals
4. ORDER	Primate	Carnivore	Rodents
5. FAMILY	Hominoid	Canid	Murine
6. GENUS	*Homo*	*Canis*	*Mus*
7. SPECIES	*sapiens*	*familiaris*	*musculus*

Three animals are classified on the chart. Look down the column for humans. Look at levels 6 and 7. Can you find the scientific name, *Homo sapiens*?

Find the scientific name for dogs. Find the name for the house mouse.

Read all the way across level 1. All three are *animals*. Read across level 2. You see that all three are *chordates* (animals with backbones). Level 3 shows that all three are mammals. They all have fur and feed their young with milk. From Level 4 to 7 each animal is completely different. Humans, dogs, and the house mouse are indeed separate kinds of animals. They must each have their own name.

That is the way the system works. Remember that the classification system is based upon *SIMILAR STRUCTURES*. It is the amount of similarity in structure that puts animals together at a LEVEL of the system.

CHECK YOUR UNDERSTANDING

1. What is the highest level in the system of classification?

2. What does the first part of the scientific name mean?

3. What does the second part of the scientific name mean?

4. Explain the meaning of this sentence: "Separate structures is the key idea in classifying animals."

> **Key Idea #3:**
>
> • A large part of the animal kingdom is made up of animals with backbones. They are called vertebrates.

ANIMALS WITH BACKBONES

Many animals are similar because they have an internal skeleton made of *BONES*. Human beings have a skeleton, and so do dogs, cats, rats, bats, and elephants. At mealtimes we often find bones from chicken, turkey, or fish. These bones have come from the internal skeleton of the animal. The house mouse also has an internal skeleton.

Vertebrates

The most important part of an animal's skeleton is the backbone. The backbone is made up of a set of bones called *VERTEBRAE*. The group of animals with vertebrae are called *VERTEBRATES*. The vertebrates include the most advanced forms of life on earth.

There are more than 40,000 species of animals that have backbones. You have already studied some of the species of vertebrates: *Homo sapiens*, *Mus musculus*, *Canis familiaris*, and *Balaeniceps rex*.

Most of the animals at the zoo are vertebrates. The giraffe, one of the stars at the zoo, is a vertebrate. *REPTILES*, another main attraction at the zoo, include lizards, snakes, turtles, alligators, and crocodiles. They all have backbones and are called vertebrates.

Giraffe

Giraffa camelopardalis

FISH belong to another large group of vertebrates. So do *AMPHIBIANS* like the toads, frogs, and salamanders. All *BIRDS* have a spinal cord made up of vertebrae. They, too, are called vertebrates.

Classes of Vertebrates

The vertebrates are not all alike, even though they all have internal skeletons made of bones. Vertebrates fall into *five separate classes*. Fish, amphibians, reptiles, birds, and mammals all belong.

The chart below shows how the leopard frog and the box turtle are classified. At the left you see the seven levels. Look at level 2. Find the word *chordate* two times. *Chordate* means that the animal has a backbone. Both the leopard frog and the box turtle have backbones. Therefore, both animals are vertebrates.

CLASSIFICATION OF TWO ANIMALS

LEVELS	Leopard Frog	Box Turtle
1. KINGDOM	Animal	Animal
2. PHYLUM	Chordate	Chordate
3. CLASS	Amphibian	Reptile
4. ORDER	Salientia*	Chelonia**
5. FAMILY	Ranid	Emydid
6. GENUS	*Rana*	*Terrapene*
7. SPECIES	*pipiens*	*carolina*

Salientia means "to leap, bound, or jump."

**Chelonia* means "tortoise" or "turtle."

Level 3 tells you that the leopard frog is an amphibian, and that the box turtle is a reptile. Levels 6 and 7 give the scientific names. The scientific name of the leopard frog is *Rana pipiens*. The box turtle's scientific name is *Terrapene carolina*.

Rana pipiens

Terrapene carolina

Animals are classified by their similarities, but the system lets you see that each kind of animal is different enough to have its own name.

CHECK YOUR UNDERSTANDING

1. What is the major similarity of structure in the animals known as the vertebrates?

2. About how many species of animals with backbones are there?

3. What are the five classes of vertebrates?

4. What is the genus of the leopard frog?

5. What is the species of the leopard frog?

> **Key Idea #4:**
>
> • Animals without backbones make up a major group in the animal kingdom. They are called invertebrates.

ANIMALS WITHOUT BACKBONES

The animal kingdom is divided into two large groups — animals with backbones, the vertebrates; and animals without backbones, the *INVERTEBRATES*.

The invertebrates outnumber the vertebrates by far. There are about 864,000 species without backbones. There are so many differences among the invertebrates that scientists have classified them into eight separate groups. Each of these groups has a large number of species. For example, in the group known as the *SPONGES*, there are 4,200 species.

ROLL CALL OF THE INVERTEBRATES

Here is a roll call and a very brief description of the groups of animals known as *INVERTEBRATES*. These animals do not have backbones.

1. Sponges

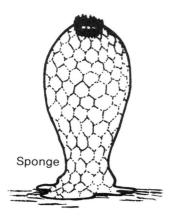

Sponge

Sponges are the simplest invertebrates. They have only two layers of cells in their body wall. They do not even look like animals. Sponges live mostly in salt water. They are attached to one spot much like plants are anchored to one spot. You may be familiar with the bath sponge, which is really the "skeleton" of a sponge.

2. Jellyfish, Sea "Flowers," and Corals

There are nearly 10,000 species of these invertebrates in the world. Most species live in the sea. These animals have tentacles with stinging cells. As they float along, they "trap" their food. Some live alone, but others like the Portuguese man-of-war live in colonies. These organisms float about for the most part. The corals, however, stay in one place. A fresh-water relative of the jellyfish is the hydra.

3. Flatworms

Flatworms have flat bodies. Some flatworms are parasites. They live inside some animal host. The most famous parasites are the tapeworm and the liver fluke. Tapeworms live as parasites in the human food canal. They live the life of "highway robbers," taking food that does not belong to them. To be free of tapeworms, humans should eat only well-cooked meat.

A fresh-water flatworm, the planaria, is interesting. If it is cut into two or more pieces, each piece will grow into a new planaria.

Planaria

4. Roundworms

There are more than 12,000 species of these tiny animals in the world. The hookworm is found in soil and water, especially where sanitation is poor. Humans may pick up hookworms through their bare feet in unsanitary places. The roundworm known as the vinegar eel can be found in cider vinegar.

Hookworm

5. Segmented Worms

There are about 3,500 species of worms with sections, or segments, in their bodies. These worms include the common earthworm (night crawler) and the leeches. Earthworms perform the useful job of tunneling their way through soil and bringing materials to the surface. The burrows or tunnels that they make let air get into the soil. Leeches are best known as blood-sucking parasites. They were once used by doctors to draw blood from patients.

6. Starfish and Other Spiny-Skinned Animals

The 5,000 species in this marine group include sea urchins, sand dollars, and sea cucumbers. Starfish are the most well-known. Starfish have tube feet on the underside of their arms that act as suction cups to hold tight to clam shells. Starfish force the clam shells open, invert their stomach into the shell, and eat the clams. Some common names are misleading. The *starfish,* for example, is no fish at all.

7. Mollusks

There are 70,000 species of mollusks, including the octopus, squid, clam, and snail. They live in the sea, on land, and in fresh water. Land snails, pond snails, and garden snails are the largest number of species in the mollusk group. Most species have shells and soft bodies. Shells are either in one piece, like the land snails, or in two halves, like the clam shell. Clams are called *BIVALVES* because of their two shells. Most mollusks have a muscular foot that they use to move around.

8. Arthropods

Arthropods are by far the largest group of invertebrates. There are more species of arthropods than there are any other group of animals in the world. Over 750,000 species are known, and more are counted each year. The arthropods are given much attention because of their great variety, their large number of species, and their special high level of interest to humans. The word *arthropod* comes from the Greek language to give a special descriptive meaning. *Arthro-* means "joint." The word *-pod* means "foot." Put them together and you have "joint foot." There are five classes of arthropods: insects, arachnids, crustaceans, centipedes, and millipedes.

Similar Structures of the Arthropods. The similar structures among the arthropods are: 1) jointed legs, 2) claws with joints, and 3) sectioned or jointed bodies. The arthropods also have *EXOSKELETONS*, or outer body coverings. The outer covering varies from *very hard,* as in crabs, to *hard,* as in certain insects and millipedes.

The *HARDSHELL CRAB* is a good example of all of these similar structures. Its legs, claws, and body parts are all in sections that are jointed. Crabs shed their exoskeletons as they grow. Each time, they grow a new exoskeleton until they become adult crabs. Arthropods follow the same general pattern. A familiar sight is the shell, or exoskeleton, of a cicada, or locust, hanging on the bark of a tree.

Hardshell crab

The Insects. More than 650,000 kinds of insects have been identified. Some are wingless, but most have one or two pairs of wings. All insects have a *HEAD* part, a *THORAX*, and an *ABDOMEN*.

Insects have a pair of *ANTENNAS* on their head. The thorax, or middle part, of their three-part body, has three pairs of legs attached. One or two pairs of wings are also attached to the thorax. The insects are a very successful group of animals. They occupy just about every kind of living space that there is: in and on the water, in trees and on plants, on the ground and under the surface, in old logs, and in most of the places where humans live. Many are pests. They can destroy crops. Some insects, however, serve as friends of humans by eating other insect pests. Insects live everywhere except in the oceans.

Ladybird beetle   Mosquito

Arachnids and Crustaceans. The most important relatives of the insects are the *ARACHNIDS* and the *CRUSTACEANS*. The arachnids include the spiders, daddy-long-legs, and horseshoe crabs. The horseshoe crab is not a crab at all. It is more closely related to the spider group. The crustaceans are the crabs, lobsters, barnacles, sowbugs, and crayfish.

Lobster   Horseshoe crab

Centipedes and Millipedes. The story of the arthropods would not be complete without including the *centipedes* and the *millipedes*. The centipedes have *one pair* of jointed legs on each of their body sections, or *segments*. They are *predators* of insects. Predators are animals that capture other animals as their main source of food. Centipedes "prey upon" insects. The millipedes have *two* jointed pairs of legs on each of their body sections, or segments. A common name for the millipede is "thousand legger." Millipedes are important forest *decomposers*. By breaking down the larger materials that fall on the forest floor, they make the soil more fertile.

Centipede

Millipede

ARTHROPODS — ANIMALS WITH JOINTED LEGS
The highest number of species on the face of the earth.

Classification and Number	Legs	Examples of Members
Insects 650,000 species	3 pairs of legs	crickets, ants, bees, houseflies, grasshoppers, butterflies
Arachnids 15,000 species	4 pairs of legs	spiders, mites, horseshoe "crab," ticks
Crustaceans 25,000 species	5 pairs of legs	lobsters, crabs, shrimp, crayfish
Centipedes 800 species	1 pair of legs per section	house centipede, garden centipede
Millipedes "Thousand-leggers" 6,000 species	2 pairs of legs per section	garden millipede meadow millipede

CHECK YOUR UNDERSTANDING

1. What are the two main "similar structures" of the group known as arthropods?
2. What is the meaning of the term *arthropod*?
3. What are the five classes of arthropods?
4. Which of the arthropod groups has the largest number of known species? How many?
5. What is the major difference between millipedes and centipedes?

SUMMARY OF CHAPTER 3

Scientists use a system for organizing animals. There are seven levels of classification. Animals are classified according to their similar structures. When an animal is classified, it is given a scientific name of two words.

Two major groups of animals in the world are the *vertebrates*, animals with backbones, and the *invertebrates*, animals without backbones. There are eight different groups of invertebrates. The largest group of invertebrates is the arthropods. Arthropods have jointed legs.

WORDS TO USE

1. wading bird
2. classification
3. feathers
4. fur
5. mammals
6. scales
7. reptiles
8. genus
9. species
10. animal kingdom
11. backbone
12. similar structure
13. chordate
14. vertebrate
15. classes
16. sponges
17. invertebrate
18. flatworm
19. jellyfish
20. segmented
21. starfish
22. roundworm
23. mollusk
24. arthropods

• REVIEW QUESTIONS FOR CHAPTER 3 •

1. What makes the shoebill stork an animal that is "one of a kind"?

2. What is the "similar structure" that is used to classify birds of the world into a single group?

3. Raccoons, tigers, cats, dogs, and giraffes are classified into a large group. What is the "similar structure" that is shared by these animals?

4. Why don't scientists use scales to separate fish from reptiles?

5. *Canis familiaris* is the scientific name for the animal you know as the dog. Which part of the name is the genus? Which part is the species?

6. What are the seven levels used in the system of classification?

7. What are some animals that have an internal skeleton?

8. How do animals qualify as "vertebrates"?

9. What are the five different classes of vertebrates?

10. What is one "similar structure" for each class of vertebrates? Write a different structure for each class.

11. What is the chief "similar structure" of the invertebrates?

12. What are the eight groups of invertebrates?

13. What group of invertebrates has the largest number of species?

14. What are the five classes of arthropods?

15. What are the two main "similar structures" found among the arthropods?

Chapter 4

Organizing Plants

Chapter Goals:

To understand that plants are classified according to their similar structures.

To know how to use the two-word scientific names of plants.

Key Ideas:

- Plants are classified into different groups, according to their similar structures.

- Seed plants include flowering and non-flowering plants with vascular systems.

- The ferns are vascular plants with roots, stems, and leaves. Ferns use spores as a means of reproduction.

- Mosses are non-vascular plants that reproduce by means of spores.

INTRODUCTION

Two parts of the world of living things are the animal kingdom and the plant kingdom. You might be tempted to ask, "Which kingdom is more important?"

The answer would have to be, "The plant kingdom." Plants could survive without animals. But animals could not survive without plants. Plants have the ability to make their own food. Animals, however, need plants for their food.

Key Idea #1:

- Plants are classified into different groups, according to their similar structures.

CLASSIFYING PLANTS

The same kind of *two-name system* that you used for the classification of animals is useful in classifying plants. Plants are classified according to the presence or absence of certain body parts, such as seeds, conducting tubes, and roots, stems, and leaves. The three main groups of plants are the *SEED PLANTS,* the *FERNS,* and the *MOSSES.*

Botanists

The system of plant classification was invented in 1753 by a man from Sweden. Carolus Linnaeus gave names to all of the known plants and animals in Europe. There are some scientists who have spent their entire adult lives studying and classifying plants. These scientists, called *BOTANISTS,* have traveled around the world in search of new species of plants. By describing the structures of plants and the places where they are found, botanists contribute to the science of *BOTANY.* Botany is one of the major branches of the science of biology.

Plants with Transport Systems

The seed plants and ferns are the major groups of plants that have an *internal system* for transporting food and water. They are called *VASCULAR PLANTS*. Vascular plants have conducting tubes called *vascular tissue*. Vascular tissue is made of *tube-like structures*. The tubes are connected into a network from the roots to the stems and into the leaves. Vascular tissue helps plants by providing a pathway for water and food to be moved to and from different parts of the plant body. For example, the roots of plants pick up water and minerals from the soil and transport them to other parts of the plant.

The *vascular system* is important in two ways. It allows the plant to grow tall. It also allows plants to grow in places where water is not always present.

Plants without Transport Tubes

In contrast to the vascular plants, there is a large group known as the *NON-VASCULAR PLANTS*. These non-vascular plants must be in *constant contact* with moisture. Because they do not have tubes for transport of food and water, they do not grow very tall. These small plants grow in damp places on the ground and on the sides of trees and rocks. They are the *MOSSES*.

CHECK YOUR UNDERSTANDING

1. What are the three major groups of plants?
2. Why are green plants so important to animals?
3. Name some "similar structures" of plants.
4. What are the differences between vascular and non-vascular plants?
5. Why is Linnaeus important in the field of biology?
6. What is the name of the special branch of biology that deals with the study of plants?

> **Key Idea #2:**
>
> - Seed plants include flowering and non-flowering plants with vascular systems.

THE SEED PLANTS

There are over 270,000 species of *SEED PLANTS*. The seed plants are the largest group of plant species in the world. Seeds contain specialized reproductive bodies which make these plants more successful than ferns or mosses.

Seed plants range in size from the smallest flowering plant, the duckweed, to the largest plants in the world, the redwoods, sequoias, and oak trees.

Seed plants have great variety and live almost everywhere. Grass, trees, garden flowers, roadside weeds, pine trees, bushes and vines of all kinds, flower-shop flowers, and cactuses are all seed plants. Also, oats, corn, wheat, rice, cotton, and barley are all plants that grow from seeds. The seed plants have vascular tissues. They are divided into two sub-groups — flowering plants, and non-flowering plants.

Flowering Plants

There are about 250,000 species of flowering plants. The word used to describe this large group is *ANGIO-SPERM*. This word was made from the Greek words *angeion*, "capsule," and *sperma*, "seed." Seeds have "seed coats" that act like small containers or capsules. Inside of the seed capsule is the stuff to begin a new plant. Angiosperms come in two kinds of seed forms called *dicots* and *monocots*. Flowering plants are one or the other.

Dicots. *Dicots* are one kind of flowering plant. There are more than 165,000 species of dicots. They have *two* "seed leaves" or *embryo* leaves inside the two halves of the seed coat. Bean seeds have two halves, and so do peanuts. You are probably most familiar with the two halves of an individual peanut.

The lima bean is a dicot plant that shows the structure clearly. The seed has two halves. When the seed is opened, you can see the embryo leaves. Also, when you grow bean plants from the seeds, you will see that the leaves have a *network* of veins rather than a set of parallel veins.

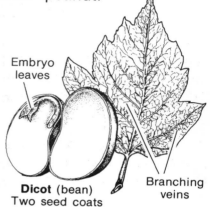

Embryo leaves

Dicot (bean)
Two seed coats

Branching veins

Some other dicots are trees that produce wood for furniture and trees that are used as shade trees. Humans use dicots in the form of fruits and vegetables.

Monocots. The *monocots* are another kind of flowering plant. There are more than 60,000 species of monocots. They include corn, wheat, and rice. Grasses that cattle feed upon are also monocots. Monocot seeds have only *one*, *single* embryo leaf. Look in a grain of corn for an embryo leaf. Monocots have leaves with veins that run *parallel*. You can see parallel veins in a blade of grass, for example.

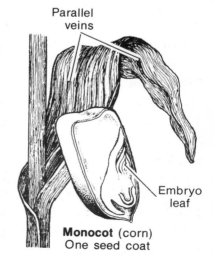

Parallel veins

Embryo leaf

Monocot (corn)
One seed coat

Non-Flowering Plants

The non-flowering plants are called *GYMNO-SPERMS*. There are about 700 species. Seeds in the gymnosperms are described as "naked seeds" because they are not enclosed in a fruit such as the apple. Instead, gymnosperm seeds are produced inside of cones. The major group of gymnosperms are the *CONIFERS*. *Conifer* means "cone-bearing."

Conifers. The conifers make up 600 species of the species of gymnosperms. Included are the pines, spruces, and fir trees. The giant redwoods and the tall growing spruces are among the conifers that make up a major part of the forests. Christmas trees are most often spruces, firs, and pines. Decorative plants such as juniper, yews, and spruces are familiar plants in the landscape of many homes.

Conifers have *needles*. The needles are the leaves of the conifers. They are able *to hold water* more easily than broad-leafed trees. This makes it easier for them to live in dry and sandy places where the tree is forced to store water for longer periods. Conifers have leaves all year round. For this reason they are called *EVERGREENS*. They lose only a part of their leaves at any time. They appear to be green all of the time.

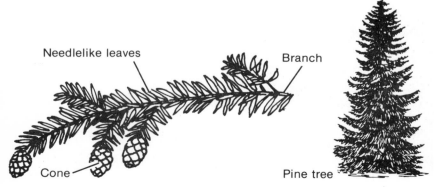

Needlelike leaves

Branch

Cone

Pine tree

Conifers occupy a large number of spots that are too harsh for other plants: along the seashores, in dry areas, and in shallow, rocky soil. You can see the bald cypress in swamps. The wood of the conifers is called "softwood." It is used as a source of lumber and pulpwood for making paper.

The Ginkgo Tree. There are other gymnosperms. The *GINK-GO TREE* of city streets is one of the most familiar. Ginkgo trees have peculiar fan-shaped leaves and are resistant to disease as well as insects. They line many city streets, as they are able to survive pollution better than other trees.

Leaves of ginkgo tree

The gymnosperms are classified according to their similar structures. They have seeds, and most have cones and needle-like leaves. They all have vascular tissue.

CHECK YOUR UNDERSTANDING

1. What are the scientific names for the flowering plants and the non-flowering plants?

2. What is a seed coat?

3. What is the difference between the dicots and the monocots? Name two plants that are dicots and two that are monocots.

4. Why are the seeds of conifers called "naked seeds"?

5. What is the shape of most conifer leaves?

6. How do these special leaves help the conifer?

> **Key Idea #3:**
>
> • The ferns are vascular plants with roots, stems, and leaves. Ferns use spores as a means of reproduction.

THE FERNS

There are about 10,000 species of ferns in the world. Many of them are tropical plants. Ferns are the largest group of *vascular plants* that *do not have seeds*. Ferns have *SPORES*. Spores are the reproductive bodies of ferns. They are found as small spots on the underside of fern leaves. Spores are formed into small clusters that dry up and split open when they are ripe. There are millions and millions of them, and they are scattered far and wide by the wind.

Spores are very, very small and are not as well equipped as seeds. They must find a moist place to land and get started. If they don't (and millions upon millions don't), they dry up. Seeds, on the other hand, carry moisture and food inside of their outer seed coats. The protective seed coat allows them to stay alive for a period of time. The seed plants are more successful for this reason.

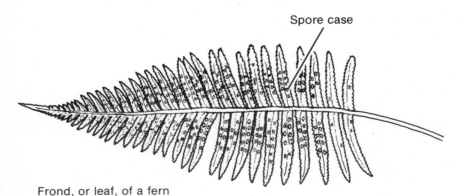

Spore case

Frond, or leaf, of a fern

Fossil Ferns

At one time the ferns were the main form of plant life on the face of the earth. Some 200,000 million years ago they were the size of trees. They became the coal deposits that are mined in different parts of the world today.

Ferns as Decoration

Ferns are used as decorative plants because of their leaves. The leaves are large and flat, and the patterns of the leaflets are attractive. Probably most of the ferns that you have seen have been potted ones. Rhizoids or runners are underground along with the roots. A stem-like midrib supports the leaflets. They are like the seed plants in having roots, stems, and leaves.

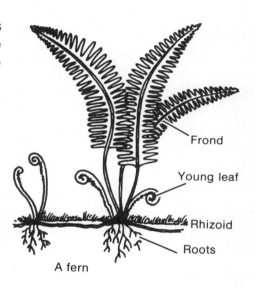

Frond

Young leaf

Rhizoid

Roots

A fern

CHECK YOUR UNDERSTANDING

1. How are ferns and seed plants alike?

2. How are ferns and seed plants different?

3. How do spores function to reproduce new plants?

4. Why are the seed plants more successful than the ferns?

5. What became of the large fern plants of long ago?

Key Idea #4:

- Mosses are non-vascular plants that reproduce by means of spores.

THE MOSSES

There are about 14,000 species of mosses. Mosses are *NON-VASCULAR* plants. They have no tubes to transport water. Therefore, they must live in moist places.

A close look at a moss reveals leaf-like and stem-like parts. They look like little trees. They often form carpet-like mats where the soil and the surrounding air are moist. Woodlands and the edges of streams are common homes for mosses. Air full of moisture, and soil that is wet provide the best home, or *HABITAT*, for the moss plants.

Moss Reproduction

Mosses reproduce by means of spores. Millions of tiny spores form inside of spore cases on special stalks or "stems." When it is ripe, the spore case breaks open. It shoots the moss spores into the air for great distances. The spores make new plants when they fall on some moist soil.

Reproduction in the mosses depends on the presence of water. The reproductive sperms need to "swim" to the egg. The resulting embryo grows into a stalk that has the spore case at the end.

Shaggy moss

Not all of the spores find good places to grow. Those that do make new moss plants. One reason for moss survival is that great numbers of spores are produced. Some spores have been known to survive for many years and to go on to produce good moss plants.

Mosses Build Soil and Bogs

Mosses have been called "pigmy plants" because of their small size and resemblance to tree plants. They are able to start growing in places where other plants cannot take root. Thin soils, cracks in rocks, and within the bark of fallen trees are natural place for mosses to grow. Mosses help in the formation of soil. When moss plants die, they form *HUMUS*. Humus helps make good soil.

Mosses of the genus *sphagnum* are known as bog builders. A bog is wet, spongy ground that results from decayed moss and other plant matter. *PEAT BOGS* are made from decayed sphagnum moss. Peat moss is used to mix with soil to improve gardens. Peat is used as a fuel in Europe.

CHECK YOUR UNDERSTANDING

1. What does "non-vascular" mean?

2. Because mosses are non-vascular, where do they need to live?

3. How do mosses reproduce?

4. How are mosses useful to humans?

SUMMARY OF CHAPTER 4

Plants are organized into three major groups, based upon their similar structures, or body parts. The three groups are: the seed plants, the ferns, and the mosses.

Seed plants and ferns are vascular plants, with tubes that transport water and food from the roots to the other parts of the plants. Of the two groups of vascular plants, seed plants are the more numerous. Seed plants reproduce by means of their seeds.

Most of the seed plants are angiosperms, or flowering plants. Angiosperms are either monocots with one embryo leaf, or dicots with two seed leaves. The majority of plants are angiosperms.

The other seed plants are gymnosperms, or nonflowering plants. Gymnosperm seeds are not enclosed in a fruit. Many gymnosperms are conifers.

Another group of vascular plants is the fern group. Although ferns have vascular tissues, they do not have seeds. Instead, they reproduce by means of spores.

Mosses are non-vascular plants. Like ferns, they reproduce by spores. Mosses help in building soil and bogs. They also provide humus which enriches the soil.

WORDS TO USE

1. transport tubes	9. reproduction	17. ginkgo
2. botanist	10. dicot	18. softwood
3. botany	11. monocot	19. needle
4. vascular	12. parallel	20. fern
5. non-vascular	13. seed	21. spore
6. Linnaeus	14. embryo	22. moss
7. angiosperm	15. conifer	23. habitat
8. gymnosperm	16. evergreen	24. bog

• REVIEW QUESTIONS FOR CHAPTER 4 •

1. Why are plants more important than animals?

2. What groups of plants reproduce with spores instead of seeds?

3. What are some of the "body parts" used to classify plants?

4. How does vascular tissue help plants?

5. What is botany?

6. What is the largest group of plant species in the world?

7. What are the scientific names for flowering and non-flowering plants?

8. What are the similar structures of dicots and monocots?

9. Why are conifers able to live in dry places? Name four trees that are conifers.

10. Why are conifers called evergreens?

11. Why is the ginkgo tree used to decorate city streets?

12. How is fern reproduction different from reproduction in seed plants? Explain how ferns reproduce.

13. Why do mosses need moisture to stay alive?

14. What is the best habitat for mosses?

15. How do mosses reproduce?

16. What are bogs? From what materials are bogs made?

Chapter 5

Organizing Protists and Fungi

Chapter Goals:

To describe the basic life activities of the protists and the fungi.

To recognize the great variety of simple forms of life in the kingdom of protists and the kingdom of fungi.

Key Ideas:

- Protists are simple plant-like and animal-like organisms that belong to a third kingdom, separate from the plants and animals.

- Protists carry out the basic life activities and affect the lives of other organisms.

- Fungi are a group of organisms that belong to a fourth kingdom, separate from plants, animals, and protists.

- Fungi carry out the basic life activities and affect the lives of other organisms.

INTRODUCTION

For a long time, keeping track of organisms was based upon the answer to the question, "Is it a plant or is it an animal?"

Some organisms have been discovered that do not fit neatly into the answer to this question. As far back as 1866, a scientist suggested that another kingdom called *Protista* should be formed to take care of organisms with special traits.

PROTISTS have special traits. They are microscopic organisms such as bacteria or algae, and they can be both plant and animal-like.

Another kingdom called *FUNGI* was added later. Fungi are organisms such as yeasts or molds. They are rooted to one spot the way plants are, but they do not make their own food.

In keeping with current knowledge, the old categories of *plant* and *animal* have been expanded to four kingdoms. In this textbook, you will study the four kingdoms of *plants*, *animals*, *protists*, and *fungi*.

Protists and fungi are classified and organized under the same two-name system that you used for plants and animals. Among all of the fungi, there is JUST ONE that is called by the name *Penicillium notatum*. In the world of the protists there is just one, and only one, *Euglena gracilis*.

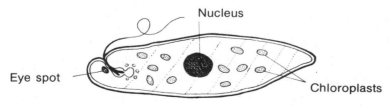

A protist, *Euglena gracilis*

```
┌─────────────────────────────────────────────┐
│  Key Idea #1:                                 │
│                                               │
│  •  Protists are simple plant-like and animal-│
│     like organisms that belong to a third     │
│     kingdom, separate from plants and animals.│
│                                               │
└─────────────────────────────────────────────┘
```

THE PROTIST KINGDOM

The protists are both plant-like and animal-like organisms. Plant-like protists such as bacteria, algae, and diatoms are able to make their own food. Animal-like protists, like the ameba, paramecium, and euglena, move around and capture food. The euglena is also plant-like because it can make some of its own food.

Plant-Like Protists

Life on earth would not be the same without the *plant-like protists*. They make their own food and are food for thousands of animal species in the seas. They are microscopic, and live in both salt and fresh water. The plant-like protists float and swim near the surface of the oceans. They reproduce rapidly. Their numbers are measured in the billions and trillions. One gallon of sea water taken from the surface would hold thousands and thousands of protists. The protists provide the *basic food source* for all of the animals of the seas.

When plant-like protists make food, they carry out the process known as *PHOTOSYNTHESIS*. During photosynthesis, the organisms give off oxygen to the atmosphere. The oceans of the world cover three-fourths of the earth's surface and are the home of the *greatest oxygen producers* — the protists. It has been estimated that these microscopic organisms provide 50% of the earth's atmospheric oxygen.

Two big jobs done by billions of these little-known workers are feeding ocean animals, and providing oxygen for other life forms.

Algae

The algae are mostly *AQUATIC* protists. They live in water. There are more than 15,000 species of algae.

1. Spirogyra

SPIROGYRA is the green algae of ponds. It is commonly known as pond scum. This bright green protist reproduces rapidly and forms floating masses on pond surfaces. There are more than 60 different species of spirogyra. All of the different spirogyra have a strand, or string, of *CHLOROPLASTS*. This green material contains *CHLOROPHYLL*, which is useful in the energy-producing process of photosynthesis.

2. Protococcus

Another genus of green algae, *PROTOCOCCUS*, is very common. Protococcus is a land organism, unlike most of the other green algae. Protococcus is found growing on the bark of trees. It grows on the shaded side of the tree. It also grows in moist places like damp stones and damp fence posts. The next time you walk in the woods, see if you can find this green algae on the shaded and moist side of the tree trunks. If you can identify that side as the north side, you have a natural compass for all other directions.

3. Sea Lettuce

Another alga that you may be familiar with is seen at the seashore. It is called *SEA LETTUCE* because it resembles regular lettuce leaves. Common garden salad lettuce is the genus *Lactuca*, and because of the strong resemblance in appearance, sea lettuce has the scientific name, *Ulva lactuca*. Notice *lactuca* in the names for both organisms. Sea lettuce may grow to six inches wide and two feet long. Sometimes after a storm, the sea lettuce fouls the shore and causes a bad odor. Sea lettuce is found along the coasts of the Atlantic and Pacific Oceans.

4. Brown Algae

The *GIANT KELP* of the Pacific coast is reported to be the longest plant in the world. This huge alga grows up to 1,500 feet and is supported by bladders to help it in floating. The plant body contains iodine and a substance called *algin,* which is used to make foods firm and uniform. This plant is harvested in California, much the same as a corn crop is harvested in the mid-west.

5. Diatoms

DIATOMS are the most numerous in the 6,500 species of brown or gold algae. They are easy to find in both fresh and sea water. They have a great variety of shapes. The shells of these microscopic organisms are made of *SILICON,* the same material that glass is made of. You can say that diatoms live in houses with glass walls!

Diatoms grow in large numbers, and they are a major source of food for many fish. The shells form silica deposits on the ocean floor. Since diatoms reproduce in a matter of days, each new reproduction period results in piles of these deposits. The "ooze" at the bottom of the sea is made of their remains. This ooze is mined to make metal polish, toothpaste, soaps, and scouring powders.

Animal-Like Protists

Animal-like protists can move on their own, and they cannot make their own food. For these reasons, they resemble animals rather than plants.

1. Ameba

The most familiar of the protists is the *AMEBA*. The way that the ameba moves is its main feature. The ameba makes false feet by flowing along. *CYTOPLASM*, or watery insides, pushes out the cell membrane. Amebas live on surfaces of rocks and plants in ponds. The ameba is just one cell. It belongs to a group known as *PROTOZOANS*.

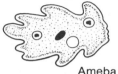
Ameba

Proto- means "first." Zoa means "animal." Their two-word scientific name is *Ameba proteus*.

2. Paramecium

Probably the second most popular organism in the world of biology is the *PARAMECIUM*. Parameciums move around to get their food and to respond to trouble in their environment. Because they move by means of *CILIA*, they are called *ciliates*. Cilia are hair-like structures.

Paramecium caudatum is a common pond inhabitant. It is more organized than the ameba. It is usually seen moving along front end first, but it does have a "reverse gear," and can back up.

3. Trypanosoma

TRYPANOSOMA is an animal-like protist that moves. It is a parasite that lives part of its life in the tsetse fly in Africa. It causes the disease, African sleeping sickness.

4. Sporozoans

Sporozoans are animal-like protists. They are all parasites and have no means of moving about. They reproduce by means of *SPORES*. One sporozoan, *Plasmodium vivax*, is famous because it is the cause of malaria in humans. This sporozoan is transferred to humans by the bite of a mosquito called *Anopheles*. Malaria is a disease of tropical countries.

Sporozoans

5. Euglena

Euglena gracilis is almost evenly divided between being plant-like and animal-like. It is plant-like because it has *chloroplasts* and *chlorophyll*. It is able to make its own food. Euglena is much like an animal because it uses a whip-like flagellum to move through the water. Euglena is a one-celled organism and fits the description of the protozoa. It has only a cell membrane like other animals; it does not have the typical cell wall that plants have. Euglena has a regular shape, but it is very flexible. It can form different shapes at times.

CHECK YOUR UNDERSTANDING

1. What are the four kingdoms used in this book?
2. What are three features of plant-like protists?
3. What are the names of five plant-like protists?
4. What do the plant-like protists do for the animals of the sea and the world?
5. What part do chloroplasts and chlorophyll play in the life of the plant-like protists?
6. What are the names of two animal-like protists?
7. How do amebas move? How do parameciums move?
8. What makes the euglena an animal-like protist?

Key Idea #2:

- Protists carry out the basic life activities and affect the lives of other organisms.

BASIC LIFE ACTIVITIES OF PROTISTS

Getting Food

The ameba, paramecium, and the euglena are microscopic organisms that feed on other microscopic organisms. Each of these protists has a special way of getting food.

The ameba moves about by the flow of its body. As it glides along, its body bumps into algae and other protists that serve as food. The ameba wraps itself around the "food." The food is trapped inside of the ameba, and it is digested. The method of wrapping around food material is a special kind of *INGESTION*.

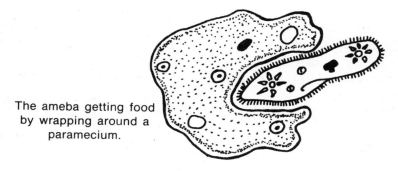

The ameba getting food by wrapping around a paramecium.

Parameciums glide through the water with the aid of their hair-like cilia. The cilia cover the body of the paramecium. They act like little oars, pushing, pulling, and turning the slipper-shaped paramecium through the water. The paramecium has a front end, a back end, and an opening in one side of its body. This opening is called a *GULLET*. As the paramecium goes along, food materials are pushed into the gullet by the cilia.

Ingested food passes to the end of the gullet and into a bubble-like space called a *VACUOLE*. Food vacuoles are the place for digestion of food. The vacuoles move around inside the paramecium's one-cell body, bringing energy to the right places.

The euglena gets most of its energy by making its own food.

Digestion and Excretion in the Protists

Amebas, parameciums, and euglenas digest their food in special places called vacuoles. Enzymes break down the food materials. The vacuoles then *EXCRETE* the waste matter.

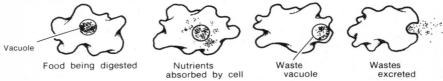

Vacuole			
Food being digested	Nutrients absorbed by cell	Waste vacuole	Wastes excreted

These protists live in a water environment. To keep a balance between the water inside them and the outside water, they have a simple *EXCRETION* system. Bubbles called *vacuoles* collect water inside, and then contract to force the water out of the protist's body. The covering, or "skin," of these simple organisms is built to let certain materials come inside. The covering also lets certain materials escape to the outside. There must be some kind kind of balance between the inside and outside so that the organism can stay alive.

Diffusion

You may wonder about water moving *into* the body of the protist. Because there are materials other than water on the inside of a paramecium's body, there is less water inside. The greater amounts of water on the outside of the paramecium tend to move to where there is less water on the inside. This process is a common event in living things. It is called *DIFFUSION*.

Diffusion is the movement of particles from one place to another. Particles of water or other substances that are more plentiful in one place will move to nearby places where they are less abundant. As materials move into and out of cells, they follow this pattern of diffusion.

Circulation in the Protists

The ameba, paramecium, and the euglena are protists with just one cell. Circulation of the cell materials is done by water moving into and out of the cell membrane. The water taken in is full of food and dissolved oxygen. Protists depend on oxygen to stay alive. The oxygen acts on food to release energy.

Coordination in the Protists

The protists react to conditions in their water environment. Temperature changes cause them to slow down or speed up their movement. A pinpoint of light will cause some of them to gather together in response. Euglena have an eye spot that is sensitive to light. They move to light and carry on photosynthesis.

Temperature and light are called *STIMULI*. The action of the protist is called a *RESPONSE*. Stimulus and response is a general feature of cells alone and of cells acting together in tissues and organs. The protists are simple organisms, but the stuff that they are made of is just like the cell materials of higher animals. All parts of their cell body are coordinated in reactions to stimuli.

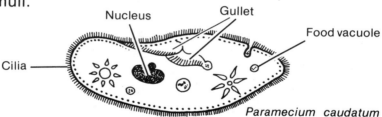

Nucleus   Gullet   Food vacuole
Cilia
Paramecium caudatum

Reproduction in the Protists

Most of the protists reproduce *ASEXUALLY*. These single-celled organisms simply divide into two *DAUGHTER CELLS*. Each daughter cell receives one half of the nucleus material. For that to be able to happen, the total materials that go to make up the two daughter cells have to be *doubled*. The process by which the nucleus divides is called *MITOSIS*. A daughter cell is like a "chip off the old block." It has the same amount of material in the nucleus as the original protist.

A paramecium dividing

Diatoms live in the warm surface waters of the oceans, where conditions are favorable for reproduction. Diatoms reproduce in a matter of hours. They reproduce asexually most of the time. They are also able to reproduce sexually, at which time the *GAMETES*, or sex cells, have a flagellum. Sexual reproduction in the protists, as well as in other organisms, requires gametes that are able to move about.

Sporozoan protists are parasites that reproduce both asexually and sexually. They have complicated life cycles in at least two host organisms. They reproduce by spores at one time and by making gametes at another.

BACTERIA

BACTERIA may be classified with the protists. They are known to be among the oldest forms of life on the earth. There are about 5,000 species of bacteria. They grow just about everywhere and are the most numerous of the different forms of life.

The Shape of Bacteria

Bacteria are classified mostly by their shapes. Bacteria may be spheres, clusters, rod-shaped, or in chains. Bacteria are measured in the smallest metric units, *microns*. It takes a thousand microns to make one millimeter. A millimeter is probably the smallest unit that you have ever worked with.

Basic Life Activities

Bacteria carry out the basic life activities. Bacteria feed on dead materials. They break down the materials that they land on. They use some of it for their own energy. Most of the broken down materials are returned to the soil and reused in the life cycles of plants and animals.

Respiration in Bacteria

Most bacteria use oxygen and carry out respiration. Bacteria are unusual, however, because some of them *do not* use oxygen. Those that do are acting in the same way as more complicated organisms. They use oxygen to release energy.

Reproduction in Bacteria

Reproduction in bacteria is by a simple division into two. Some bacteria divide every twenty minutes. Their numbers are terrific! One common bacteria is *Escherichia coli*, and it divides quickly and often. You recognize the two-name system at work for even the smallest of organisms. Scientists shorten *Escherichia* to *E.* and call this organism *E. coli*. It is found in the digestive tract of human beings.

Bacteria Are Spoilers

Some bacteria cause food to spoil. For example, when bacteria from the air invade milk, the milk sours. Bacteria grow their fastest in warm places. Refrigerating food reduces the chances of spoilage.

Helpful Bacteria

Bacteria are *useful* in several ways. They help in the making of several major foods, such as butter, yogurt, and cheese. Bacteria are the organisms used in modern genetics. They are small, reproduce rapidly, and populations of millions and billions are useful in the study of changes that take place in genetic material. Bacteria are also useful in *GENETIC ENGINEERING*. This fairly new science allows scientists to grow organisms with specific abilities. Bacteria have been engineered to digest waste. Oil spills may be "eaten" by specially designed bacteria that have the ability to digest and change oil.

Nitrogen and Bacteria

Nitrogen is the most plentiful element in the air, but plants are not able to use it directly. Bacteria work in the soil and around the roots of plants. Certain bacteria act on the nitrogen in the air and make it into chemical compounds that plants are able to use. This action of bacteria in the soil may be the most important thing that bacteria do. Bacteria also act on chemical remains of living materials to break them down. In that way the same nitrogen gets back into the atmosphere. Once back in the air, it is ready to be recycled through the soil bacteria into plants.

CHECK YOUR UNDERSTANDING

1. How does the ameba ingest food?
2. What are vacuoles? How do they work?
3. What is diffusion? How does it work?
4. Why is circulation so simple in the protists?
5. What is a stimulus? What is a response?
6. What is the main way that protists reproduce?
7. What is the most important thing that bacteria do?

Key Idea #3:

- Fungi are a group of organisms that belong to a fourth kingdom, separate from plants, animals, and protists.

THE FUNGI KINGDOM

Why is it necessary to have a separate kingdom for fungi? Fungi are different from plants. Fungi do not make their own food. They have no chloroplasts or chlorophyll. Fungi do not have leaves, stems, or real roots with root hairs. Fungi are different from animals. They do not move from place to place. They are non-vascular organisms that reproduce by means of spores. Fungi are grouped alone in the fungi kingdom.

Life Activities of Fungi

Fungi grow in areas of rotting leaves and wood or decaying animal matter. Fungi are the great *DECOMPOSERS* of the world. They produce substances called *ENZYMES*. The enzymes dissolve surrounding material that the fungus absorbs or takes in as food. Fungi are important in the *cycle* that involves breaking down waste matter such as dead leaves, rotting wood, and decaying animal life. The dissolved chemicals help other organisms. Trees can take in the chemicals from the soil by way of their roots.

The main growth of fungi is a mass of interconnected tubes. The greater part of a fungus is below the surface on which it is growing. What you see above the surface is the soft, fruiting body. Fungi do not grow tall or make any woody tissue.

Mushrooms

The best known of the fungi are the *MUSHROOMS*. Mushrooms grow in damp and dark places. The forest floor is a common habitat, or home. Mushrooms come in all colors. Although white and brown are the most common, you can also find purple, red, and yellow mushrooms. What you see and what people eat is the fruiting body. Spores are made in the umbrella-shaped part. Most of the mushroom is found beneath the surface. Underground is a mass of what looks like white threads all strung out.

Mushrooms are a delicacy. In France and Italy people eat dark mushrooms known as truffles. Truffles grow beneath the ground and are rooted out by specially trained pigs.

Mushrooms

Yeast

YEAST qualifies as a fungus because it lacks chlorophyll. It also uses spores for reproduction the way that other fungi do. Yeast also reproduces by *BUDDING*. Yeast spores are everywhere. They are carried by air currents.

Yeast is used in baking to make the bread rise. Yeast cells cause *FERMENTATION* in carbohydrates. Enzymes from yeast break down the carbohydrates into carbon dioxide and alcohol.

In grapes the carbohydrates are sugar. When yeast cells mix with the sugar, the grapes become wine. Beer is produced with the help of yeast cells. German brewers brought their *yeast cultures* with them when they came to America. They continued the German tradition of beer making in this country.

Yeast cells

Yeast cells are often studied in biology classes because they are easy to grow in sugar water. They are easy to find under the microscope. Their method of forming buds that grow into new cells is very plain to see. A *population* of yeast cells lives, grows, and dies off in a short period of time. Learning just how a population grows is a part of the study of biology.

Molds

Most people are familiar with *MOLD*. Bread gets moldy, and so does jelly, fruit, meat, and cheese. Even in the cold of the refrigerator things get moldy. Mold grows best in moist and dark places. Mold grows on different materials, such as wood, cloth, leather, paper, and on many living things.

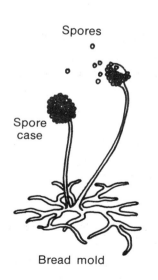

Spores

Spore case

Bread mold

Bread Mold. The spores of the mold *Rhizopus* are in the air almost everywhere. Bread can get moldy with no trouble at all. Set a slice of moist bread on a saucer, and spores will land on it. As the spores grow, they send threads of mold into the bread. The mold makes digestive enzymes that dissolve the "food" in the bread for its own use. Some of the bread you buy has preservatives in it to keep the mold away. The name of the black bread mold is *Rhizopus nigricans*.

Penicillin. Some molds are not pests at all; they are helpful. The genus *Penicillium* is the mold that is effective in attacking and reducing bacteria. Alexander Fleming was the scientist who discovered penicillin. Quite by chance he observed that the mold was killing off bacteria in the material that he had been working with. This discovery led to the use of penicillin as a medicine for killing microorganisms that are harmful to humans.

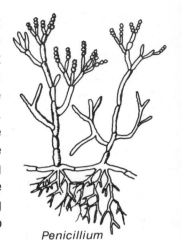

Penicillium

Molds and Cheeses. The flavor of different kinds of cheeses is caused by different species of molds. *Penicillium roquefortii* is the name of the mold that provides the flavor for roquefort cheese.

Rusts and Smuts

Rusts and smuts are fungi that are *parasites* that live on crop plants. They are destructive parasites because they can kill the *HOST* plant. A host plant is one on which parasites feed and grow. These parasites cause millions of dollars of damage to crop plants every year.

RUSTS are reddish brown fungi that grow on plants like wheat, corn, barley, and rye. The plant turns the color of rust.

A *SMUT* is a fungus that grows on cereal crops. Smut is noted for its foul smell. This fungus changes plant organs into dark masses of spores. It is most familiar on corn plants.

Some Famous Fungi

In addition to penicillin, another famous fungus is the one that infected the potato crop in Ireland in the middle of the 1800's. This *BLIGHT* fungus caused a serious famine in Ireland. Many people died of starvation. As a result of the blight on the potato crop, many people left Ireland and came to settle in the United States.

Another famous fungus is the one that causes "athlete's foot." This fungus infection of the feet is widespread. It causes discomfort, but it is not deadly. Fungi that *can* be deadly, however, are wild mushrooms. They are famous because they can poison people.

RINGWORM is a fungus infection of the human scalp and skin. It grows in ring patterns. The "worm" part of "ringworm" is another example of how common names can be confusing; the cause of ringworm is not a worm at all. It is caused by a fungus.

Lichens

Lichens are unusual because they are really two organisms growing together. Lichens are fungi and algae living together in a partnership. This partnership is called *mutualism* or *symbiosis* because the fungi and the algae depend upon each other. Just as in most partnerships, each partner has a job to do. In the lichen partnership the fungus provides moisture because of its ability to absorb and hold water. The alga has chlorophyll and is able to make food that the fungus can use.

Lichens are food for lower animals. One famous lichen is the "reindeer moss" which provides food for caribou herds, a kind of northern reindeer.

Lichens can grow in places where other organisms cannot survive. For this reason they are often the first to grow in a new area, and are sometimes called "pioneer" organisms. Lichens are able to withstand great extremes of temperature and are often found in very cold places, such as the North and South Poles.

Lichens can live nearly everywhere. They are not tall. They live close to the surface where they attach. Lichens can be found on the barks of trees, on the surface of bare rocks, and on barren mountain tops. Different species with a variety of colors often "paint" rocks into striking natural patterns. Lichens are very sensitive to pollution and are not usually found in cities.

CHECK YOUR UNDERSTANDING

1. What makes it necessary for fungi to have a kingdom of their own?

2. What is the name of one of the best-known fungi?

3. What are three useful things that yeasts do?

4. What is budding? How does it work?

5. What is the genus and species of the black bread mold?

6. Where does penicillin come from? What does it do?

7. Where would you look for rusts and smuts?

8. What are three famous fungi? Why are they famous?

9. What is so unusual about the lichens?

Key Idea #4:

- Fungi carry out the basic life activities and affect the lives of other organisms.

FUNGUS LIFE ACTIVITIES

How Fungi Get Food

Fungi get their food by *ABSORPTION*. Whatever they are living on or in is the source of food for the fungi. Spores that land on bread send runners down into the bread. The runners give off digestive enzymes. These chemicals dissolve the bread. The chemicals of the bread become the *energy source* from which the fungi live, grow, and reproduce.

Because fungi lack chlorophyll, they must depend entirely on either green plants or on animals that have green plants at the beginning of their food chain. *BRACKET FUNGI* grow on the sides of trees. They look like little house roofs. They depend on the tree for their food and energy. *MILDEWS* grow on old leather, paper, cloth, and anything else that is a product of living things. Mildews depend on moisture and their digestive enzymes to break down the chemicals in these products for use as food.

Fungi and Roots

Some fungi live in a close relationship with the roots of green plants. The fungi take water from the soil and hold it for use by the plant. Minerals are also absorbed by the fungi. The green plant provides the fungi with nutrients which it dissolves with its digestive enzymes. There are some forest trees that do not grow as well as they might if the fungi are absent from the soil around their roots.

The fungi and green plant relationship is of mutual benefit. This is called *MUTUALISM*, a special kind of living together where both members benefit.

The Great Decomposers

Fungi are the world's great *DECOMPOSERS*. Fungi may not look very important, but they are essential to all other living things. The chief role of the fungi is to decompose, or break apart. Fungi secrete digest-ive juices into the material that they are living on — leaves, soil, rotting logs, animal remains, jelly, jam, fruit. They digest the material for their own use, but most of it is returned in broken-down form to the soil to be re-used by other organisms.

Fungi decompose all kinds of waste products, remains of plants, and remains of animals. They help recycle useful chemicals and substances. What would the world be like if the fungi were not on the job, recycling all the dead remains?

Reproduction in the Fungi

Fungi reproduce *asexually* by means of *SPORES*. Spores are reproductive cells. They are very small and are produced in tremendous numbers by each fungus.

Bread Mold. Bread mold reproduces by two methods. One method is *asexually* when a single spore *GERMINATES*, or begins the growth pattern of a new cycle that ends with the making of new spores. The second method is *sexually*, when gametes, or sex cells, join together from two strains of mold cells. The body that is formed by sexual reproduction germinates and produces a new set of spores. Bread mold reproduces rapidly. A whole loaf of bread that the mold lands on gets covered *overnight*!

Yeasts. Yeasts reproduce by *BUDDING*. A yeast cell grows to its regular size, and a swelling begins on the cell wall. The swelling grows and then pinches off from the "parent" yeast. The new yeast grows to its regular size, grows some buds of its own, and the process continues. Yeast populations feed on sugar and grape juice and grow very rapidly to large numbers.

The growth of a population of yeast can be followed by watching and counting over a period of days. Using the microscope as a tool for studying reproduction and population growth in yeast will give you the flavor of how scientists go about doing their work.

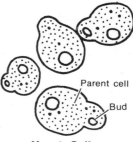

Yeast Cells

Growth of a Fungus

A fungus grows in moist, dark, and warm places. Growth is the result of a set of tube-like threads that mesh in with soil, bread, leaves, or fruits of all kinds. Wherever there are carbohydrates, the fungi grow. They form a mass of mostly white and often fuzzy looking threads. Inside of the threads are many nucleuses, but no cell walls. Materials for growth are transported in this fragile and moist mass of threads.

Over 100,000 Species

Specialists in the study of fungi have identified over one hundred thousand species of fungi. Fungi grow everywhere that you can think of. Their spores are in the air you breathe, in the waters of the world, and in every nook and cranny of the world of soil. You already know two of them that live on human bodies — the fungus of athlete's foot and the one that causes ringworm.

CHECK YOUR UNDERSTANDING

1. How do the fungi get their food?

2. What is mutualism?

3. Why are the fungi called the "great decomposers"?

4. What is the main method of reproduction in the fungi?

5. How do yeasts reproduce?

6. Where are some of the places that the fungi grow?

SUMMARY OF CHAPTER 5

Protists and fungi are so different from plants and animals that they are classified separately into their own kingdoms.

Protists are simple animal-like and plant-like organisms. Protists play important parts in the lives of other organisms. Chlorophyll in the protist, green algae, is important in the food-making process. Protists reproduce asexually. Many of the protists use mitosis and cell division to produce the next generation. Protists do all of the same kinds of things that other organisms do.

Fungi are different because they lack chlorophyll and live on other organisms or materials. Fungi are especially important as decomposers. Fungi are responsible for decomposing, or breaking down, dead animals and plants and putting this changed matter back into the soil, so it can be re-used by other organisms. Fungi reproduce asexually. Some fungi produce different strains that form gametes and reproduce by sexual means.

Bacteria are protists. They are the most numerous of organisms on earth. There are many useful bacteria that help in food processing. Bacteria can be harmful by causing disease and spoiling food. The most important thing that bacteria do is to fix the *NITROGEN* of the air into useful chemicals for plants to use in making proteins.

WORDS TO USE

1. protists
2. fungi
3. photosynthesis
4. algae
5. aquatic
6. spirogyra
7. chloroplast
8. chlorophyll
9. protococcus
10. sea lettuce
11. brown algae
12. diatoms
13. silicon
14. ameba
15. protozoans
16. paramecium
17. cilia
18. trypanosoma
19. sporozoans
20. spores
21. euglena
22. ingestion
23. gullet
24. vacuole
25. excretion
26. diffusion
27. stimuli
28. response
29. asexually
30. daughter cells
31. mitosis
32. gamete
33. bacteria
34. nitrogen
35. genetic engineering
36. decomposer
37. enzymes
38. mushrooms
39. yeasts
40. budding
41. fermentation
42. molds
43. penicillin
44. parasites
45. host
46. rusts
47. smuts
48. blight
49. ringworm
50. lichens
51. mutualism
52. absorption
53. bracket fungi
54. mildews

• REVIEW QUESTIONS FOR CHAPTER 5 •

1. Why is there a separate kingdom for protists?
2. What part do the plant-like protists of the oceans play in the amount of oxygen in the air?
3. What is the main difference between spirogyra and protococcus?
4. What is the connection between sea lettuce and regular lettuce?
5. What organism lives in a "house with glass walls"?
6. What organism is a ciliate? Tell how it moves.
7. What is the cause of the disease known as malaria?
8. Why is *Euglena* so hard to put into one group?
9. What are two ways that the animal-like protists get their food?
10. How does diffusion work in the paramecium?
11. What is the purpose of the euglena's eye spot?
12. What is mitosis?
13. What are the two ways that diatoms reproduce?
14. What are two ways in which people use bacteria?
15. What do the enzymes of the fungi do?
16. What is fermentation?
17. What is the relationship between a parasite and a host plant?
18. In the lichen partnership, how do the algae and the fungi help each other?
19. Where do mildews grow? What do they use for food?
20. Why are fungi called the great decomposers?
21. Describe the process of budding in yeast cells.

Chapter 6

Discovering How Animals Stay Alive

Chapter Goal:

To discover how animals use their body systems to carry out the basic life activities in order to stay alive.

Key Ideas:

- Animals use a variety of food getting and food ingesting methods to maintain their bodies.

- Digesting food and getting rid of waste products are necessary to help animals stay alive.

- Systems for circulation and respiration provide food and oxygen to all animals' cells.

- Coordination of animal activities is done by a nerve network or nervous system.

INTRODUCTION

The key to staying alive is to provide for the body. Supporting and maintaining the cells, tissues, organs, and organ systems is a full time job. Some of the activities that support life are getting food, moving, growing, using food, and removing waste. How animals stay alive is the story of how they carry out the basic life activities.

> **Key Idea #1:**
>
> • Animals use a variety of food getting and food ingesting methods to maintain their bodies.

FOOD GETTING AND INGESTING

You can understand more clearly some of the ways animals get food by studying five different animals.

Methods of Getting Food in the Sponges

Sponges, the simplest of animals, have a system for pumping water in and out of their bodies. In the process of pumping, sponges take in diatoms and other small organisms for food. This food is their source of energy. Sponges are attached to one spot on the ocean floor. They are known in biology as *SESSILE* animals: animals that are attached to one spot by some kind of *holdfast cells*. Since they are not able to move around to get food, they must continually pump lots of water every hour of every day.

Because of the very small size of the animals, plants, and protists that the sponges take in for food, it is necessary for huge numbers of organisms to be "eaten" by the sponge.   The sponge just holds tight to its place and pumps and pumps.   Getting food means bringing in many, many gallons of water filled with tiny organisms.   Food getting in sponges is a simple plan for a simple animal.

Food Getting in Hydras and Jellyfish

Hydras are sessile animals for most of their life. They attach by a holdfast, stretch their bodies and tentacles to great lengths, and sting and capture small water animals.   Hydras are freshwater inhabitants of ponds and streams.   They are predators that use their tentacles to bring their *PREY* into a central opening or mouth.   *Prey* organisms are animals taken by other ani-mals as food.   Food is ingested by the movement of the hydra's tentacles.

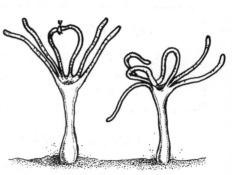

A hydra ingesting its prey

Jellyfish get their food as they float in their marine environment with their tentacles streaming along beneath them.   Fish that run into the tentacles are stung and captured.   The tentacles bring the food to the central part of the jellyfish, where it is ingested.

Jellyfish also take in microscopic green algae that float about in the sea.   These algae are members of the protist kingdom.   They are important producers of food and oxygen in the oceans of the world.   There is a great variety of these algae.   All together, these algae are called *PLANKTON.*   The plankton of the seas is the main source of food and energy for larger organisms to use to stay alive.

How Spiders Get Food

The nest or the web of the spider is the special feature of its life. Getting food is the name of the game. Spiders build elaborate nests and webs to capture insects. Once entrapped, the insect prey is wrapped and the blood sucked from its body.

A spider uses its web to get food.

How Insects Get Food

Some insects eat plants by using *CUTTING* jaws on their mouthparts. *Walking sticks* chew on tree and plant leaves. *Termites* eat on wood, and may eat away the whole foundation of a building. They have *CHEWING* mouthparts. The *ant lion*, or doodle-bug, feeds on ants that are caught in sliding sand pits. Adult *dragonflies* eat insects which they capture on the wing. These beautiful insects also eat mosquito larvae.

The largest group of insects in the world are the *beetles*. One of them is called the confused flour beetle, and it has *CHEWING* mouthparts to eat flour, grain, and anything starchy like beans or peas. The flour beetle is often found in the house with the cereal or cornmeal. Another member of this group of beetles is the *Japanese beetle*, which is a pest that feeds on the leaves of grapevines, apple trees, cherry trees, corn, roses, blackberries, strawberries, and hollyhocks.

One whole group of insects are called *bugs*. They have *SUCKING* mouthparts and include the squash bug, bedbug, backswimmer, and water boatman.

Butterflies have *SUCKING* mouthparts to siphon the nectar from flowers. The larvae of butterflies are caterpillars that chew the leaves of many plants. The periodic *locust* has a sucking beak to take juices from roots and stems.

The *horsefly* and the *deerfly* have *PIERCING* mouthparts and live on the blood of warm-blooded mammals. The common housefly feeds on dead and decaying matter. The housefly has the scientific name *Musca domestica* and is not to be confused with the horsefly.

How Frogs and Toads Get Food

Frogs and toads are amphibians and live much of their lives around water or in moist places. They eat insects, earthworms, and other small animals. The tongue of these animals is attached to the floor of the mouth. It lashes out with a flicking type motion, and in an instant, faster than the human eye can see, it captures a flying insect. The end of the frog's tongue is sticky, and the insect has little chance to escape. Capturing and ingesting food in frogs and toads is a highly special act. It helps them stay alive.

Frogs and toads capture prey with their tongues.

The bullfrog, *Rana catasbeiana*, is a large frog with a diet of insects, other frogs, fish, and even birds. Bullfrogs are famous for their great jumping skills. Many fine restaurants serve the legs of these frogs as a delicacy.

Another famous amphibian is the American toad, *Bufo americanus*. *Bufo* lashes out its sticky tongue at ants, slugs, earthworms, beetles, and bugs in general. *Bufo* is a frequent inhabitant of back yards.

CHECK YOUR UNDERSTANDING

1. What are the scientific words for breathing, eating, and drinking?

2. What is the main part of the story about how animals stay alive?

3. What do sponges do to get food?

4. What is similar about how a hydra and jellyfish get their food?

5. What is the special feature of a spider's life?

6. What does the doodle-bug feed on?

7. What is the method for getting food among the beetles?

8. What is the difference in the food for the horsefly and the common housefly?

9. What is so unusual about the way that frogs and toads get their food?

> **Key Idea #2:**
>
> • Digesting food and getting rid of waste products are necessary for animals to stay alive.

DIGESTION AND EXCRETION

DIGESTION is a basic life activity in which food is broken down into small particles and then into simpler chemicals so that it can be used. Special chemicals called *ENZYMES* mix with and dissolve food.

EXCRETION is a basic life activity in which *WASTE MATTER* is removed from cells or from animal bodies. Solid, liquid, and gaseous materials are removed by some form of *EXCRETORY SYSTEM*.

Digestion and Excretion in Sponges

Even simple organisms like the sponges have enzymes in their body cells. The enzymes are chemicals that act upon food that came in along with the water being pumped through the body wall of the sponge. The body wall of the sponges is only two cell layers thick, and yet the whole process of digestion takes place there. Enzyme action digests the food, and cells use the energy to keep the animal alive.

Energy is released from the food and is used to carry out the basic life activities. Carbon dioxide and other waste products are formed. The sponge gives off the waste materials through cells in the body wall. The simple organization of the sponge body makes digestion and excretion direct and simple processes.

Digestion and Excretion in Hydras

Hydras have tentacles that are used to capture food in the form of small water animals. The food is brought to the mouth opening and into a hollow body sac. This is where *DIGESTION* takes place. The cells lining this sac give off digestive juices called enzymes.

The food stuff is reduced to usable small chemicals that go to the cells. Cells use the food for energy and release waste products into the body sac. Water and carbon dioxide are the main waste materials formed from cell respiration. They are *EXCRETED* through the mouth opening.

Digestion in Spiders

Digestion in spiders is unusual. It takes place outside of the body of the spider for the most part. Spiders feed on insects. They feed on the soft inner parts. To do this, spiders use a pair of special fangs near the mouth. The fangs first inject poison to kill the prey. Then holes are punched to allow digestive juices from the spider's mouth to enter the prey's body.

The digestive juices turn the soft inside parts of the prey into a fluid. Spiders take in only *liquid food*. The fluid is sucked up into the spider's body and into its stomach. Further digestion takes place. Food molecules are absorbed by the blood.

Digestion in Insects

One example of an insect is the grasshopper. Grasshoppers eat grass and have mouth parts that chew. The digestive system of the grasshopper is a tube that begins with the mouth. The mouth is connected to the esophagus, which in turn, is connected to the crop. The crop is a storage place and leads to the gizzard, where food is ground up. The ground-up food goes through to the *stomach* where enzymes digest the food. The food molecules are absorbed through the stomach wall into the bloodstream. After digestion, some solid wastes move from the stomach through the *intestine* and out of the grasshopper through the *anus*.

Digestion and Excretion in Frogs and Toads

Frogs and toads are great insect eaters. They are valued by humans because they help control the populations of insects. They have large, wide mouths that help in the capture of insects.

The digestive system of these two amphibians is similar to that of the vertebrates in general. In fact, frogs are studied in some detail in biology courses because they can be compared to the human. They have the same organs of digestion, the same kinds of digestive glands, and a similar blood system.

Food that enters the mouth is pushed into an opening called the gullet. Then it moves to the esophagus, which is a tube that connects to the stomach. Not much happens to the food until it reaches the stomach, where digestive juices in the form of enzymes begin to break the food apart. The next stop is the *SMALL INTESTINE*, where bile from the *LIVER* and digestive enzymes from the *PANCREAS* act some more on the food. Frogs and toads and other vertebrates have a *GALL BLADDER* that stores and releases the bile that breaks down *FATS*.

The food mass that is formed by digestion is now in a usable chemical form. The *BLOOD VESSELS* that surround and are imbedded in the small intestine wall absorb the food material and circulate it to all parts of the body. The cells use the food for energy to move muscles and to carry out the basic life activities in general.

There are some waste materials that are left after digestion of the main part of the food. These semi-solid products are moved on to the *LARGE INTESTINE*, and from there to the *CLOACA*, and out of the body through a small opening called the *ANUS*.

Meanwhile, back in the body cavity, the *KIDNEYS* are filtering waste products from cell respiration from the blood. This liquid waste is called *URINE*, and is stored temporarily in the *BLADDER* before being released to the *CLOACA*.

The gaseous waste product of cell respiration is *CARBON DIOXIDE*, and it is released and eliminated through the skin, lungs, and mouth.

The digestive system of the vertebrates is also called the *ALIMENTARY CANAL*. It is a system of organs that maintain the life of the organism. It is closely associated with the excretory system and with the circulatory system.

CHECK YOUR UNDERSTANDING

1. What is the meaning of *digestion*?
2. What is *excretion*?
3. What is the method of digestion in the sponges?
4. How do hydras digest their food?
5. What makes digestion in the spiders so unusual?
6. What is the job of the gizzard in the digestive system of insects?
7. What happens to food in the small intestine of the frog?
8. What is the job of the kidneys in the excretory system of the frog?

Key Idea #3:

- Systems for circulation and respiration provide food and oxygen to all animals' cells.

CIRCULATION AND RESPIRATION

Organisms must get food, oxygen, and water to their body cells. Animals use a pump and a *CIRCULA- TORY SYSTEM* of conducting tubes, or vessels, to do that job. The pump is a heart, and the vessels are arteries and veins.

Organisms must get oxygen to their body cells. A breathing or *RESPIRATORY SYSTEM* of some kind does that job. Animals rely on a system of windpipes, lungs, and blood vessels to get oxygen to all of the right places.

Circulation and Respiration in Hydras and Jellyfish

The cells of hydras are in direct contact with the water environment. Circulation of water carries materials into and out of the cells. Inside the hydra's body, food materials are digested and circulated directly into the cells that line the inside wall. Waste materials that result from cell respiration — water and carbon dioxide — diffuse from the cells to the body cavity and are expelled through the mouth opening. The body plans for circulation and respiration in the hydras and jellyfish are simple compared to the verte- brates.

Circulation and Respiration in Insects

Insects are very active organisms. Flying, running, and jumping require energy. The circulatory system is an open one. There is a heart in the upper part of the body cavity. It is not a complicated heart, just a swelling of a tube called the *aorta*.

Blood of insects is a watery fluid. Blood moves food, oxygen, and carbon dioxide around the body parts. Insects have a special feature in their respiratory system: tubes connect the outside air with the inside body. The openings of the small tubes run along the sides of the insect body and open and close in rhythm. The abdomen moves somewhat like an accordion. Air is forced in and out quite rapidly to take care of the needs of the active insect.

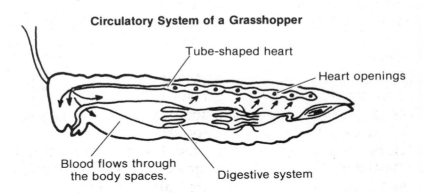

Circulatory System of a Grasshopper

Tube-shaped heart

Heart openings

Blood flows through the body spaces.

Digestive system

Circulation and Respiration in Spiders

Spiders have a set of "book" lungs located in the body cavity. This is in addition to their air tubes. The book lungs get their name from their appearance of pages in a book with spaces between them. The air spaces are connected to the outside air. Blood circulates in the "pages," and plenty of air is in contact with the blood. Exchange of oxygen takes place. The oxygen is used in cell respiration in the spider's cells.

Circulation and Respiration in Frogs

The circulatory system in the amphibians is much more complex and highly organized than in the sponges, hydras, jellyfish, insects, and spiders. The problems are bigger. In frogs there is much cell activity deep inside the body. There are many more layers of cells. The cells are not exposed in the watery environment as are those of the sponges, hydras, and jellyfish. The body plan of frogs and other vertebrate animals must be able to circulate materials to a much deeper and more spread-out variety of cells.

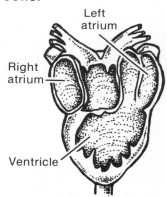

Frog Heart

The frog has three chambers in its heart. The chambers are connected to each other and to blood vessels. Blood circulates in vessels to carry oxygen to body cells. Blood picks up and gets rid of carbon dioxide. Blood vessels carry blood that is full of oxygen to the head, legs, and main body. Another set of blood vessels returns blood full of waste carbon dioxide from the body, legs, and head to the heart. The heart is the manager of what blood goes where. The chambers serve as collectors, sorters, and senders.

Within the body of frogs and other vertebrates there are two kinds of circulation. One kind works on circulation to and from the lungs — lung circulation. The other works on circulation to and from the body — body circulation. Both circulations use blood vessels to transport the blood. The heart is the pumping organ that is in the middle of the two different circulations — lung circulation and body circulation.

Each circulation forms a complete circuit. The lung circuit is from heart to lungs and from lungs to heart. The body circuit is from heart to body parts and back to the heart. *ARTERIES* are the blood vessels that take the blood *away from* the heart. *VEINS* are the blood vessels that *return* the blood to the heart. Artery blood is full of oxygen. Arteries get their oxygen supply in the lungs and through the skin of the frog. The blood takes the carbon dioxide out of the frog by way of the lungs. Some carbon dioxide goes out of the frog's body through the skin.

The purpose of the circulation of blood is to get oxygen to all parts of the body. The frog's cells use the oxygen to carry out *CELL RESPIRATION*. Food taken in by the frog goes through the digestive system and ends up in the small intestine. Blood vessels along the intestine take up the dissolved food and oxygen and transport them to the cells, where energy is released.

Frogs spend much of their time in water. They have a thin skin. They are able to take oxygen through their skin to their body cells and blood vessels. They also get rid of some waste carbon dioxide and water through their skin.

CHECK YOUR UNDERSTANDING

1. What is the circulatory system?
2. What does the respiratory system do?
3. What makes circulation and respiration so easy in hydras and jellyfish?
4. What is a special feature of the respiratory system in the insects?
5. What method do spiders use for getting oxygen?
6. What makes the circulatory system of frogs so much more complex than other organisms?
7. What are the two kinds of circulation in frogs?

> **Key Idea #4:**
>
> • Coordination of animal activities is done by a nerve network or by a nervous system.

COORDINATION

Coordination of body activities means getting the different parts to act together. Body parts must act smoothly. Coordination of body activities in the invertebrates, like sponges and hydras, is carried out by simple nerve networks. These nerve networks respond to objects and changes in their surroundings.

Coordination in multi-cellular vertebrate animals is not so simple. Frogs, birds, raccoons, horses, and humans have complicated body systems that must all work together to get things done. Muscles, organs, and bones must work smoothly for these animals to stay alive. Coordinating the work of the muscles, organs, and bones is the job of the *NERVOUS SYSTEM*.

Coordination plays a big part in all of the basic life activities, but especially in movement, sensing and reacting, and getting food.

Coordination in the Sponges

The sponges are simple animals and have no centralized or highly developed nervous system. Sponges do not get complicated problems to work on. They dwell in warm water where conditions stay much the same.

The cells of sponges carry out simple activities. Each kind of cell does a different job. Cells in sponges act alone; they do not form into tissues. Separate parts of sponges are so independent that if a sponge is cut into pieces, each piece is able to re-grow into a new sponge.

Coordination in Hydras and Jellyfish

Unlike the sponges, the more highly developed hydras and jellyfish have tentacles that must be coordinated. Hydras are attached and move their tentacles about, but they also move around. Hydras have been observed doing "tentacles over holdfasts" similar to human gymnasts doing handsprings. This is possible because of communication between nerve cells and cells that act like muscle cells.

A hydra can move by flipping over.

Hydras react to the touch of a needle or other object in their environment by pulling all parts of the body together. This happens because of a *NERVE NET* that is a kind of low level nervous system. This same nerve net is what coordinates the movement of the hydra's tentacles as it captures food and then brings the food into an opening or "mouth" at the base of the tentacles. If chemicals are put into the water, or if the temperature is altered, the body of the hydra responds by pulling its parts together.

Jellyfish pulsate and move around by means of a muscle-like ring of cells. The tentacles bring captured food into the center of the body. Coordinated food getting and movement are done by a nerve net similar to the hydra's.

Coordination in the Insects

Insects have sense organs. Eyes, antennas, and a *TYMPANIC MEMBRANE*, or ear, are all used to respond to stimuli from the environment. Reach for a grasshopper and it quickly jumps or flies away. Most insects avoid being captured. The bee diagram below shows a brain, a nerve cord running the length of the body, and a pair of eyes. Most insects have these kinds of parts in their nervous systems.

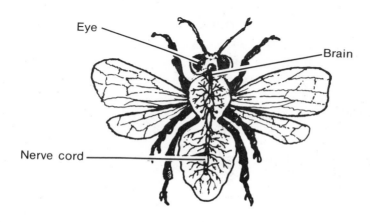

Nervous System of Honey Bee

Insects eat a variety of foods, use a variety of methods for getting food, and have a wide variety of behavior and movement patterns. All of this requires a high degree of coordination. Insects are *among the most successful* of all kinds of organisms on earth. Part of their success is due to their coordinating system. The brain of most insects has a highly developed set of *OPTIC LOBES*. This part of the brain coordinates eye movements. Muscles in the legs, in the area of mouth parts, and at the base of the wings act together in food getting or escaping from being food for some other organism. Insects have been around a long time.

Coordination in Spiders

Spiders have also been on earth for a very long time. Spiders have a nerve cord that attaches to a fairly large brain. Spiders carry on a variety of activities. Those spiders that weave webs do so by instinct. They do not have to be trained, or learn how. Web making is an inherited or inborn skill.

To make their webs, line their nests, and wrap their prey, spiders make silk. On the underside of the spider are *SPINNERETS*, organs that make silk threads.

Spiders have eight simple eyes that are connected by nervous tissues to the brain. The presence of an insect trapped in the web is sensed by sensory hairs on the legs of the spider. The tarantula does not see very well. It is covered with velvety hairs, and it relies on its sensory touch. The spider is a web builder and a predator. These activities require coordination.

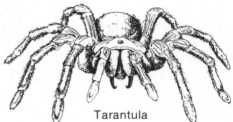
Tarantula

Coordination of Frog Activities

The *NERVOUS SYSTEM* of the frog is like that of most vertebrate animals. Biology classes study the frog's *ANATOMY* to understand how human systems look and how they work.

All of the activities of a frog's body are coordinated by the *NERVOUS SYSTEM*. The nervous system of the frog is made up of the *BRAIN*, the *SPINAL CORD*, and *NERVES*. One set of nerves runs from the brain to parts of the head. The eyes are important sense organs that are connected by nerves to the brain. All actions of the nervous system are very fast. That is a feature of nervous tissue.

The Spinal Cord. A second complete set of nerves connects the spinal cord to all parts of the body cavity, or *ABDOMEN*. The nervous system is the *COORDINAT-ING SYSTEM*. It acts very much like the conductor of a symphony orchestra. It controls the other systems of the body. It does this because all of the cells of the body have nerve endings that receive signals through the nervous system. Nerve tissue goes to all parts of an organism. A main feature of cells is that they are *SENSITIVE* to a number of stimuli. Most of the muscles in the frog's body respond to actions that occur in the brain. One set of nerves controls voluntary muscle actions by way of the brain.

Automatic Actions. A separate and third set of nerves controls all of the automatic actions of the body. The heartbeat, changes in sizes of blood vessels, and action of the digestive system are controlled by automatic actions of nervous tissues.

The Brain. The brain is the organ of the nervous system that is in charge. It is at work all of the time. Seeing, feeling, sensing, smelling, and hearing are all *COORDINATED* in the brain. Even tasting gets some action in the brain.

A set of muscles that move has received a signal from the brain on a *MOTOR NERVE*. The original signal may have started when the frog was touched. Then nerve endings in the skin sent a signal on a *SENSORY NERVE* to the brain. Motor nerves and sensory nerves are the coordinating network of verte-brate animals.

Being alive is a matter of *responding* to the large numbers of *stimuli* in the environment. Frogs hear other frogs during the mating season. They respond to the mating call of their species.

Using the Two-Name System. The common species of frog studied in biology classes is *Rana pipiens*. It is sometimes called the grass frog. There is only one frog in the two-name system that goes by that name. *No other frogs admitted!*

Frogs Are Complex. Unlike the sponges, hydras, jellyfish, and spiders, the frog is able to do a much greater number of things. Frogs can leap, flick out their tongues to capture insects, swim with great speed, make sounds like RIVET! RIVET! RIVET!, and escape the grasp of a human trying to make a capture. Frogs reproduce by sexual reproduction that is coordinated by their nervous systems.

CHECK YOUR UNDERSTANDING

1. What is the job of the nervous system?
2. What is it that makes responses in sponges?
3. What is the "nerve net" in hydras?
4. What are the main parts of the nervous system in insects?
5. What do spinnerets do for spiders?
6. What are the main parts of the nervous system in the frog?
7. What are some automatic actions controlled by nerves in the frog?
8. What is the difference between a motor nerve and a sensory nerve?

SUMMARY OF CHAPTER 6

Animals stay alive by carrying out the basic life activities. Animals have different ways of ingesting food, and they use a variety of methods for digesting food. The variety in methods runs from the simple "sitting and getting" to the complex flicking of a frog's tongue to capture a flying insect. From the stinging tentacles of jellyfish to the biting jaws of the doodlebug, the methods of getting food run the gamut.

Digestion and excretion require systems from the simple diffusion of the sponges and jellyfish to the complicated systems of the frogs and toads. Breaking down animal and plant food into useful chemicals of food by the action of enzymes helps animals stay alive. Spiders live on a liquid food diet. Insects have a tube-like digestive system with special organs that do special jobs along the route. Getting rid of solid and liquid wastes is the job of the excretory system. Carbon dioxide waste material is also given off by animals.

The circulatory system and the respiratory system work together to transport important materials in animal bodies. Digested food, oxygen, and carbon dioxide are the main passengers in the fluids of the circulatory system. Lungs and air tubes help bring in oxygen and get rid of carbon dioxide. Insects and spiders with the beginnings of real hearts pump blood through the body. Frogs and toads have three-chambered hearts that take care of lung circulation and body circulation. Frogs also take oxygen from their water environment through their skin.

Coordination of body activities is done by nerve networks and by complex nervous systems. Insects have brains, nerve cords, and connecting nerves. Frogs and toads are representative vertebrate animals with brains, spinal cords, sensory and motor nerves, and organs such as eyes, ears, and specialized mouths. Spiders build complex webs by instinct or inborn patterns of nervous behavior. The nervous system in insects is responsible for a wide variety of activities: flying, jumping, swimming, crawling, fighting, mating, egg-laying, and making sounds. Staying alive means carrying out all of the basic life activities in a coordinated and consistent way.

WORDS TO USE

1. sessile
2. predator
3. prey
4. tentacles
5. plankton
6. mouthparts
7. bladder
8. digestion
9. excretion
10. enzymes
11. esophagus
12. crop
13. gizzard
14. small intestine
15. large intestine
16. anus
17. liver
18. gall bladder
19. alimentary canal
20. cloaca
21. circulatory system
22. respiration
23. book lungs
24. lung circulation
25. body circulation
26. blood vessels
27. cell respiration
28. abdomen
29. brain
30. coordination
31. spinal cord
32. nerve net
33. nervous system
34. tympanic membrane
35. optic lobes
36. spinnerets
37. motor nerve
38. sensory nerve

• REVIEW QUESTIONS FOR CHAPTER 6 •

1. What are some of the essential things that animals must do to stay alive?

2. What is the difference between ingestion and digestion?

3. How do sponges get their food?

4. What do the hydra's tentacles do?

5. What is the main method for getting food among the spiders?

6. What are some of the different kinds of mouthparts among the insects?

7. How can you keep from confusing the horsefly and the housefly?

8. What is so special about the way that frogs and toads use their tongues?

9. What do enzymes do?

10. How do hydras get their food?

11. What purpose do the fangs on the spider serve?

12. Name the five parts of the grasshopper's digestive system.

13. What are the five parts of the digestive system in a frog?

14. What is another name for the digestive system of the vertebrates?

15. What system is used to move or transport food and oxygen in an organism's body?

• REVIEW QUESTIONS FOR CHAPTER 6 •

16. What is the main job of a respiratory system?

17. What is special about the respiratory system in insects?

18. What are book lungs? What makes them work so well?

19. What is the main job of "lung circulation" in frogs?

20. What is the main job of "body circulation" in frogs?

21. What does the skin of a frog have to do with oxygen and carbon dioxide?

22. What is the nature of the nervous system in sponges?

23. What is a nerve net?

24. What is the difference between the tympanic membrane and an optic lobe?

25. What are the main parts of the nervous system in the insects?

26. Why is the weaving of a spider's web called an instinct?

27. What are spinnerets?

28. What are the main parts of the nervous system of a frog?

29. What do motor nerves do?

30. What is the job of a sensory nerve?

Chapter **7**

Discovering How
Plants Live Their Lives

Chapter Goals:

To discover how plants live their lives by transporting food and water, by making food, and by giving off oxygen.

To understand the way in which special plant parts are used for reproduction.

Key Ideas:

- Seed plants and ferns use vascular tubes to move food, water, and minerals through their roots, stems, and leaves.

- Green plants use carbon dioxide, water, energy from the sun, and the chlorophyll in their cells to make food.

- Green plants release oxygen into the air during the food-making process.

- Plants use flowers, fruits, seeds, and spores to reproduce.

INTRODUCTION

What if you couldn't move? You wouldn't be able to walk into your favorite fast food place. If you could ever get there, all you could do would be to stand in just one spot. You couldn't even open your mouth or swallow. You would be like a plant and would need to grow some roots, stems, and leaves. You would have a system of vascular tubes instead of blood vessels. Your flowers, fruits, and seeds would be parts of your reproductive system. Or, if you were a moss, you would reproduce by spores. Instead of breathing out carbon dioxide, you would exhale oxygen. And, instead of blood cells, you would have special green cells called *CHLOROPLASTS*. The chloroplasts would enable you to make your own food. You wouldn't even need that fast food place, because you could make all the food you needed. In fact, you could feed the animals of the world. They would be dependent upon you, not only for their food, but also for their oxygen.

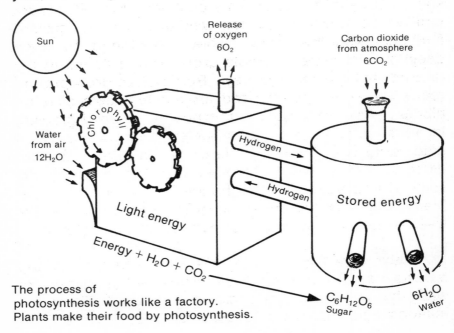

The process of photosynthesis works like a factory. Plants make their food by photosynthesis.

> **Key Idea #1:**
>
> • Ferns and seed plants use vascular tubes to move food, water, and minerals through their roots, stems, and leaves.

THE TRANSPORT SYSTEM IN VASCULAR PLANTS

Roots Move Water and Minerals

The vascular system in plants begins with the roots. Roots are out of sight. They grow under the ground. What do roots do? Roots hold the plant in one place. They take in water and minerals from the ground. They can soak up, or *ABSORB*, the water and minerals and store them until needed. Or, the roots can move the water and minerals to other places in the plant, using their *XYLEM TUBES*.

How Do Roots Work? *ROOTS* grow into and through soil. This growth takes some push or force. Roots are built to be able to push their way around in the ground and find what they need. They grow longer to reach deep ground waters. Roots grow around things in the ground that get in their way. Roots often grow around rocks under the ground and around the roots of other plants. Have you ever tried to pull out a weed growing in your yard? Were you surprised at how strongly it held onto the ground? Were you surprised at how deep or widespread the root system had grown in the soil? Roots are always on the job in the lives of plants. Many root systems under the ground are as big as the tree or bush that they support above the ground.

Parts of a Root. *ROOT HAIRS* are thin tubes that extend from the root cell wall. They are the connection with the water and minerals in the soil. There are millions and millions of such small tubes that actively soak up water and minerals from the ground water supply.

A special structure at the end of each root is the *ROOT CAP*. It protects the main part of the root as it grows and pushes its way through the soil. Root cap cells wear out as the root grows, so new root cap cells are made all of the time.

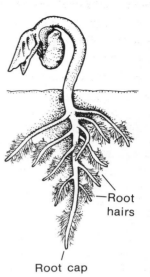

—Root hairs

Root cap

Inside the root, and *up into* the stems and leaves, *XYLEM* tubes move incoming water and minerals. Above the ground and down into the root cells, special tubes called *PHLOEM TUBES* carry food *from the leaves* to all parts of the plant.

Stems for Support

A *STEM* is the part of a plant's body that supports the leaves and connects the leaves and the roots. Stems store food and move food, water, and minerals through the plant.

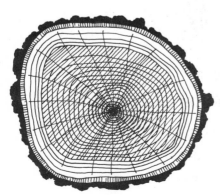

Cross section of tree with growth rings

Plants also contain a layer of tissue that produces new xylem and phloem cells. The layers are responsible for new growth. The rings that you count to tell the age of a tree are called "annual rings."

The Leaves of Plants

Leaves are mostly thin and flat with three main parts. The *PETIOLE*, the *BLADE*, and the *VEINS*. The petiole attaches the leaf to the stem or to a branch. The blade is the big part where food is made. The veins are scattered through the blade and run through the stalk to the stem. The blade is built to get plenty of light from the sun. A single leaf has very much surface area. A whole tree full of leaves, then, is able to gather huge amounts of energy from the sun.

The veins of leaves have a system of skinny tubes that are arranged in a pattern. Veins transport food and water between the main stems and the leaf. Small openings (seen with a microscope) on the underside layer of each leaf allow gases like carbon dioxide and oxygen to enter and leave the leaf. Water vapor, an invisible gas, also enters and leaves through these openings called *STOMATES*, or *PORES*. Most of the water gathered by the roots passes through the vascular system of the plant and is transferred to the air through the pores.

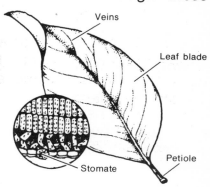

A leaf is made up of layers of cells.

CHECK YOUR UNDERSTANDING

1. What do roots do for plants?
2. What are the three main parts of a leaf?
3. What are stomates or pores?
4. What do veins do in a leaf?

> **Key Idea #2:**
>
> - Green plants use carbon dioxide, water, energy from the sun, and the chlorophyll in their cells to make food.

FOOD MAKING IN GREEN PLANTS

Special cells in the plant leaf hold a green pigment called *CHLOROPHYLL.* Chlorophyll uses the energy of sunlight to change carbon dioxide and water into *SIMPLE SUGARS* and oxygen. The simple sugars are food for the plant, but the process is not called cooking. It is called *PHOTOSYNTHESIS.*

How Photosynthesis Works

Plant leaves contain special cells called *CHLORO-PLASTS.* Chloroplasts contain chlorophyll, a green substance. When sunlight hits a leaf, the chlorophyll captures the light energy and keeps it there. Carbon dioxide comes from the air and enters the leaf through the stomates. Water comes up through the roots.

The plant uses the energy from the sun to *split up the water* into its two parts: *HYDROGEN* and *OXYGEN.* The oxygen from the water is sent into the air. Hydrogen, the other element found in water, is combined with carbon dioxide to produce simple sugars. Because of its chlorophyll, the plant is able to change *LIGHT ENERGY* from the sun into *CHEMICAL ENERGY.*

Chemical Energy from Food

What does chemical energy have to do with making the food that you eat? Simple sugars are carbohydrates. Sugar is food. The food you eat contains carbohydrates, fats, and proteins.

The process of photosynthesis is the food-making process in plants. The simple sugars made during photosynthesis combine with other molecules to make different kinds of carbohydrates, such as starches. They also combine with other molecules to form fats.

NITROGEN, a part of the air, is needed to make proteins.

How Chemicals Combine in Photosynthesis

During photosynthesis, chlorophyll breaks up water molecules into hydrogen and oxygen. Hydrogen combines with carbon dioxide to form simple sugar. The breaking up and the combining of molecules are called CHEMICAL REACTIONS. Everything in the world is made up of MOLECULES, and molecules are made of tiny ATOMS. An atom is the smallest part of a molecule. A molecule is the smallest combination of atoms which forms a separate substance. When atoms combine and recombine, different molecules are formed.

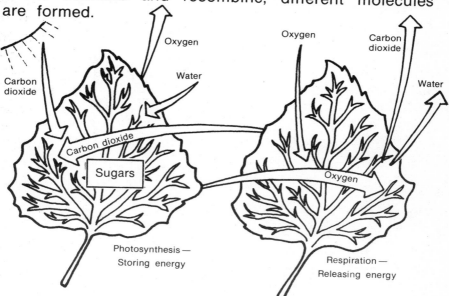

Plants use oxygen and carbon dioxide
in photosynthesis and respiration.

Chemical Formulas

The chemical formula for water is H_2O. Water is made up of two parts hydrogen and one part of oxygen. The H stands for the hydrogen atom, and the O is the symbol for the oxygen atom. The small number 2 is right next to the symbol for hydrogen. That means that there are two hydrogen atoms in a molecule of water. There is no number near the letter O. This means that there is only one oxygen atom in a molecule of water.

For 12 molecules of water, the formula for water would look like this: $12H_2O$. How many hydrogen and oxygen atoms are in 12 molecules of water? If you said 24 hydrogen atoms and 12 oxygen atoms, you would be correct. Each molecule of water contains two hydrogen atoms (H_2). Therefore, 12 molecules of water have 12 × 2, or 24 atoms of hydrogen. Each water molecule has only one oxygen atom in the formula, so 12 water molecules contain 12 × 1, or 12 oxygen atoms.

Photosynthesis

The Formula for Photosynthesis

The chemical formula for carbon dioxide is CO_2. Each molecule of carbon dioxide has one atom of carbon and two atoms of oxygen. The formula for six molecules of carbon dioxide is $6CO_2$. Six molecules of carbon dioxide contain six carbon atoms and 12 oxygen atoms. In the process of photosynthesis, 12 molecules of water and six molecules of carbon dioxide combine to form one molecule of sugar, six molecules of oxygen, and six molecules of water. The chemical equation for the entire process of photosynthesis looks like this:

$$12H_2O + 6CO_2 + \text{light energy} + \text{chlorophyll} = C_6H_{12}O_6 + 6O_2 + 6H_2O$$

$$\left[\text{water} + \text{carbon dioxide} + \text{light energy} + \text{chlorophyll} = \text{simple sugar} + \text{oxygen} + \text{water} \right]$$

In an equation, the left side and the right side are equal. In the equation for photosynthesis, there are an equal number of oxygen, hydrogen, and carbon atoms on each side.

CHECK YOUR UNDERSTANDING

1. Why are plants so important to animals?

2. What is the special material inside plant leaves that enables them to carry out the process of photosynthesis? What does this material do?

3. During photosynthesis, plants change one type of energy into another. What kind of energy does a plant get, and where does this energy come from? After photosynthesis, what is the new form of energy called?

4. What kind of food does a plant make?

5. What are the four items that are necessary in order to have photosynthesis?

6. What are the three items that are produced as a result of photosynthesis?

> **Key Idea #3:**
>
> • Green plants release oxygen into the air during the food-making process.

PLANTS GIVE OFF OXYGEN

During the food-making process in the leaves of plants, *OXYGEN* is released. Oxygen is a gas that is useful to living things.

What Is a Gas?

A *GAS* is something that you can't see or feel, but it is really there. Do you know why it is so light that you can't feel it and so clear that you can't see it? It's because the molecules in a gas are spread out and move around quickly. In a solid, such as your desk, the molecules are packed tightly together and hardly move at all. That's why the desk is so hard.

Why You Need Oxygen

The air you breathe is composed of carbon dioxide, nitrogen, and oxygen gases. Oxygen is one of the key gases in the air. Oxygen is very important in the lives of plants and animals. If oxygen is absent, energy is not released from food. Oxygen *unlocks food energy* which is stored in the cells of plant and animal bodies. This process is called *CELL RESPIRATION*.

One kind of respiration is breathing in and out. You inhale air full of oxygen and exhale air full of carbon dioxide and water. Body cells use oxygen to break apart the sugar molecules that were built in plant life. Cell respiration is a special *low temperature* kind of burning. Animals and plant cells have a special ability to "burn" food. You get energy when your cells burn up food. This is cell respiration.

How Plants Produce Oxygen

During the process of photosynthesis, plants use the chlorophyll in their special cells called chloroplasts. Using the chlorophyll and the energy from the sun, plants combine water and carbon dioxide to make simple sugar. During this process, two other substances are produced besides the simple sugar. Do you remember what they are? Look at the formula for photosynthesis on page 113. You can see that each time a plant makes one molecule of simple sugar, it also makes six molecules of oxygen and six molecules of water. The plant uses some of the oxygen for getting its own energy. The rest of the oxygen goes out the stomates into the air. The oxygen that is released to the air came from the water when it was broken down in photosynthesis.

Plants Release Oxygen

Plants put the oxygen they don't need back into the air. The oxygen exits through the many pores in the leaf. These pores, as you know, are called stomates. Each stomate has two special cells called *GUARD CELLS*. Guard cells open and close the stomate to let water and gases in and out. Extra oxygen exits from the plant when the guard cells bend apart. The opening between the guard cells is the pore that lets out the oxygen. During periods of the day, the two guard cells move back together to close the stomate, and gases are not exchanged. The opening and closing of stomates is controlled by the amount of carbon dioxide inside the guard cells.

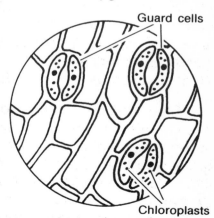

Guard cells

Chloroplasts

Stomates on Underside of Leaf

Plants *take in* carbon dioxide and *give off* oxygen during photosynthesis. Animals *breathe in* oxygen and *breathe out* carbon dioxide and water. This continuous process or cycle is called the *OXYGEN-CARBON DIOXIDE CYCLE.*

Besides green plants, the protists of the oceans go through the same oxygen-carbon dioxide cycle. Protists supply more than half of the oxygen in the world's atmosphere.

CHECK YOUR UNDERSTANDING

1. What is the difference between the molecules in a gas and a solid?

2. What is the process of cell respiration?

3. Why does a plant have extra oxygen?

4. What do guard cells do?

5. What is the oxygen-carbon dioxide cycle?

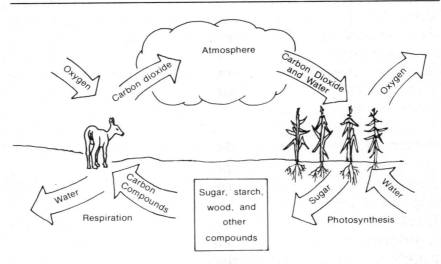

The Oxygen-Carbon Dioxide Cycle

> **Key Idea #4:**
>
> • Plants use flowers, fruit, seeds, and spores to reproduce.

REPRODUCING NEW PLANTS

Plants can reproduce by *SEXUAL REPRODUCTION* or *ASEXUAL REPRODUCTION.* The prefix "a" in the word *asexual* means "not," so the word *asexual* means "not sexual." Sexual reproduction involves two parents. One parent provides the egg cell, and the other parent provides the sperm cell. In the plant world, seed plants reproduce by sexual reproduction. In asexual reproduction, only one parent is involved. Plants with spores, such as mosses and ferns, reproduce both sexually and asexually.

Asexual Reproduction

Plant-like protists reproduce asexually. One-celled organisms divide into two cells, and each cell becomes a new organism. Simple green algae reproduce this way. Plants such as mosses and ferns reproduce by forming *SPORES.* A spore is a reproductive cell with a thick protective coating. New cells are formed when the parent cell divides. The wind can blow the spores or they can fall to new surroundings where they will develop into new plants.

Sexual Reproduction in Mosses and Ferns

Mosses and ferns reproduce both sexually and asexually. During sexual reproduction a *SPERM CELL* and an *EGG CELL* join together to form a *ZYGOTE CELL.* The sperm cell is the male cell responsible for reproduction. The egg cell is the female cell. The zygote cell is a new organism. It is the beginning of a new plant. The sperm cells swim to the egg cell through the moisture in the plant and fertilize the egg. This is called sexual reproduction, because two different sex cells or *GAMETES* take part in the process.

Reproduction in Seed Plants

Seed plants reproduce sexually. The egg cells are encased in a structure called the *OVARY*, and the sperm cells are found in *POLLEN GRAINS*. For reproduction to take place, the pollen must fertilize the egg. Insects, wind, or rain can transport the pollen to the egg.

How Flowering Plants Reproduce

The *FLOWER* is the special part of a plant's body which makes both egg cells and sperm cells. Flowering plants, or *angiosperms*, have many different kinds of flowers such as roses, tulips, and daisies. The flowers may look different, but they all contain the same kind of reproductive parts.

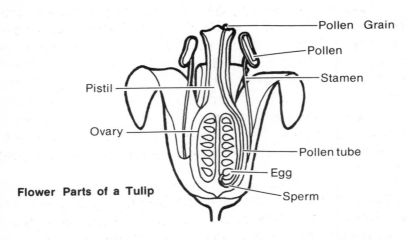

Flower Parts of a Tulip

In a flowering plant, or *angiosperm,* the *STAMENS* are the male organs of reproduction, and the *PISTILS* are the female organs. The stamens and pistils are both contained in the flower of the plant.

The first step in reproduction is *POLLINATION*. Pollen is transferred from the stamen to a sticky substance at the top of the pistil. If pollination takes place within one flower, it is called *SELF-POLLINATION*. If a flower from one plant pollinates the flower from another plant, it is called *CROSS-POLLINATION*.

Fertilization

After pollination, the pollen grain grows a tube. The *POLLEN TUBE* grows long enough to reach down to the *OVARY*, which is at the bottom of the pistil. The ovary contains the eggs. When the pollen from the pollen tube meets the egg, *FERTILIZATION* takes place. Fertilization is the beginning of reproduction. After fertilization, the ovary grows and becomes a *FRUIT* with seeds inside. The job of the fruit is to house and protect the seeds of the next generation of plants.

Seeds

SEEDS are capable of germination to produce a new plant. Seeds contain the *EMBRYO*, or beginning stages of a new plant. When temperature and moisture conditions are right, a seed begins to germinate. *GERMINATION* is the beginning of development. Seeds carry all of the basic information to make a new plant just like the ones from which it came. Pea plants, bean plants, dandelions, and grapes all come from seeds that came from pea plants, bean plants, dandelions, and grapes.

Water is the key factor in the beginning of germination. Once water has been soaked into the seed from the environment, many chemical changes begin to take place. Special chemicals called *ENZYMES* cause these changes. Enzymes start things going.

All of the activities of living things depend upon energy. Energy comes from the stored food that is the part of every seed. This food is starter food. Very soon the plant grows to the place where it can make its own food. As the plant grows, it develops all of the regular parts that you recognize — roots, stems, and leaves. Later you would recognize flowers, fruits, and a new set of seeds inside the fruit. The new seeds are at once, the end of one cycle of germination, growth, and development, and the beginning of the cycle that is the next generation.

How Conifers Reproduce

A pollen grain and an egg are the reproductive cells of gymnosperms, or non-flowering, plants. The cone of a gymnosperm is like the flower of an angiosperm. Cones contain the reproductive organs of the plant. However, there is one big difference between reproduction in conifers and in flowering plants. In flowering plants, both male and female reproductive parts are most often in the same flower. In plants with cones, some cones are male, and other cones are female. Some species have *trees* that have separate sexes. Male ginkgo trees and female ginkgo trees are one example.

First, the pollen grains are released from the *male cone*. Millions of grains are scattered about and can be carried by the wind or rain. Only a few grains reach a female cone. When they do, the pollen grain grows a tube that reaches down to the egg. When the pollen and egg meet, *fertilization* takes place. Just as with flowering plants, pollination can take place within the same tree. This is self-pollination. Or cross-pollination can occur if two trees are involved.

CHECK YOUR UNDERSTANDING

1. What is the difference between sexual and asexual reproduction?

2. What are gametes?

3. How does fertilization take place in a flowering plant?

4. What is the difference between self-pollination and cross-pollination?

5. What is the difference between the organization of reproductive parts in angiosperms and gymnosperms?

SUMMARY OF CHAPTER 7

Plants live their lives by using their body parts to survive and to reproduce. Roots and stems give plants their support. Root hairs soak up water and minerals from the ground. In vascular plants, such as ferns and seed plants, xylem and phloem tubes carry these substances through the plant.

Plants capture sunlight through their leaves. Tiny holes in a leaf, called stomates, allow materials to flow in and out. The veins in leaves have a system of tubes that transport food and water. The pattern of the veins and the shape of the leaves help identify the plant.

Plant leaves contain a green substance called chlorophyll, which enables the plant to make its own food. Food making, or photosynthesis, depends upon light energy from the sun. Plants use chlorophyll to break up water molecules into hydrogen and oxygen. The plant then combines the hydrogen with carbon dioxide from the air to produce a simple sugar. In this chemical reaction, plants change light energy from the sun to chemical energy. Without plants, there would be no food for animals.

During photosynthesis, plants produce oxygen and water in addition to simple sugar. Plants put the oxygen they don't need back into the air through the stomates in their leaves. Guard cells open and close the pores to let materials in and out. Plants take in carbon dioxide and give off oxygen. Animals inhale oxygen and exhale carbon dioxide. This exchange of gases is called the oxygen-carbon dioxide cycle.

Mosses and ferns reproduce both sexually and asexually. Spores, which are the reproductive units in these plants, can divide in half to produce new organisms. This process is called asexual reproduction, because only one parent is involved. Mosses and ferns also reproduce sexually. The two gametes, the sperm cell and the egg cell, join together to produce a zygote.

Seed plants reproduce sexually. In angiosperms, the flower contains both the male sex organ, the stamen, and the female sex organ, the pistil. First, pollen from the stamen is transferred to a sticky substance at the top of the pistil. This process is called *self-pollination*, if the male and female sex organs are in the same flower, or *cross-pollination* if two plants are involved. Next, the pollen grain grows a tube which reaches down to fertilize the ovary at the bottom of the pistil. The ovary grows and becomes a fruit with a seed inside.

Gymnosperms have cones which contain the reproductive organs. Pollen grains of male cones are scattered about by the wind or rain to fertilize the eggs of female cones. Just like flowering plants, either one or two plants may be involved in reproduction.

WORDS TO USE

1. chloroplasts
2. vascular plants
3. absorb
4. xylem tubes
5. roots
6. root hairs
7. root cap
8. phloem tubes
9. stem
10. minerals
11. woody stems
12. petiole
13. blade
14. veins
15. stomates
16. pores
17. chlorophyll
18. simple sugars
19. photosynthesis
20. oxygen
21. hydrogen
22. light energy
23. chemical energy
24. nitrogen
25. chemical reactions
26. molecules
27. atoms
28. cell respiration
29. guard cells
30. oxygen-carbon dioxide cycle
31. sexual reproduction
32. asexual reproduction
33. spores
34. sperm cell
35. egg cell
36. zygote cell
37. gametes
38. ovary
39. pollen grains
40. flower
41. stamens
42. pistils
43. pollination
44. self-pollination
45. cross-pollination
46. fertilization
47. fruit
48. seeds
49. embryo
50. germination
51. enzymes

• REVIEW QUESTIONS FOR CHAPTER 7 •

1. What do xylem and phloem tubes do?
2. What are the two important jobs of roots?
3. What are the three important jobs of stems?
4. What are the two features of leaves that help identify plants?
5. What is photosynthesis?
6. What are four things needed for photosynthesis?
7. What are three things made by photosynthesis?
8. What are stomates, and what do they do for plants?
9. What is a chemical reaction?
10. What two vital substances do plants provide for animals?
11. How many carbon atoms and how many oxygen atoms are there in the formula $6CO_2$?
12. What is a simple sugar?
13. What is cell respiration?
14. What gases do plants and animals exchange? Which organism supplies which gas?
15. What is the difference between the way molecules act in a gas and in a solid?
16. What is a zygote?
17. What is a spore?
18. What does a flower do?
19. What is the difference between pollination and fertilization?
20. What is a seed, and from where does it come?

Chapter 8

Understanding Systems of the Human Body

Chapter Goals:

To describe the structure and function of the systems of the human body.

To recognize that the systems work together in the body as a whole and carry out the basic life activities.

Key Ideas:

- Ingestion and digestion make food available and ready for use by body cells.

- The circulatory system, the respiratory system, and the excretory system transport food molecules, oxygen, and carbon dioxide.

- The nervous system, sense organs, and endocrine glands send and receive information to control and coordinate the body.

- The skeletal and muscular systems provide support, shape, and protection. They work together to make body movements.

INTRODUCTION

Imagine the best automobile in the world. It has a frame that is strong and yet is light. Its body is sturdy and smooth, so strong that it resists most dents and scratches, but smooth and beautiful to look at. Imagine that if it *does* get a scratch, it repairs itself and looks as good as new. Its gas mileage is very efficient. Its body parts are always well lubricated and smooth running. They last until the car is very old. They do not require much maintenance, only basic good care and servicing.

Imagine that carbon monoxide exits through the exhaust system without polluting the air. The automobile responds to external signals such as traffic lights and other vehicles. When its oil or battery is running low, for example, it will flash a light, so that the oil or battery can be replaced or changed.

The amazing car that you have just imagined is *not nearly as amazing as your body*. Like the imaginary automobile, your body has a frame for support and to protect the vital parts inside. This frame is made of bones, which are strong but light. When you get a cut, you bleed, but then the blood stops flowing, and eventually the skin heals itself, usually without leaving a scar.

Your brain and nerves, which comprise your nervous system, take in signals from the outside and respond by doing things such as moving body parts. Better than any automobile or machine manufactured by people, your body is the most marvelous "machine" in the world.

> **Key Idea #1:**
>
> - Ingestion and digestion make food available and ready for use by body cells.

INGESTION AND DIGESTION

Getting food into your body is *INGESTION*. Ingestion is eating and drinking. Breaking down food into pieces which are simple enough to go into the cells is called *DIGESTION*.

When digestion is over, pieces of original food are watery chemicals, so small that they cannot be seen. Even if you looked through the most powerful microscope in the school, you would not see them. If you did see something, you would not even recognize it as food. It would have been changed into chemicals called *FOOD MOLECULES*.

Digestion Begins inside Your Mouth

You chew and crush food with your teeth and jaws, turning it over with your tongue. This is called *MECHANICAL DIGESTION* because the size of the food particles is changed. When more than just the *size* of the food pieces is changed, the process is called *CHEMICAL DIGESTION*.

Chemical digestion begins in the mouth. Under your tongue are the *SALIVARY GLANDS*. These glands make a liquid chemical called *SALIVA*. Saliva changes some of the food while you chew and before you swallow. It changes the food into a simpler form.

Tongue

Salivary glands

From the Mouth to the Esophagus

Knowing what happens to the food you eat while it is in your mouth may not seem exciting. It is, however, very important to the body you live in. You probably enjoy eating. Food often tastes good, and sometimes *feels* good — like cold ice cream, a juicy apple, or crispy crackers. Taste buds in the tongue are sensitive to four tastes — sweet, sour, salty, and bitter. Chewing food and moving it around in the mouth prepares it for swallowing and movement through the *ESOPHAGUS*, a long tube that connects the mouth to the stomach.

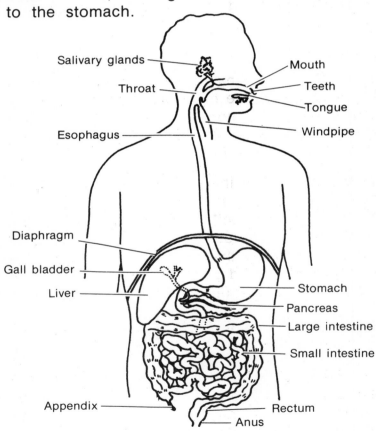

Salivary glands

Throat

Esophagus

Mouth

Teeth

Tongue

Windpipe

Diaphragm

Gall bladder

Liver

Stomach

Pancreas

Large intestine

Small intestine

Appendix

Rectum

Anus

Human Digestive System

The Stomach

Your *STOMACH* is an organ that stores the bulk of your meal. Digestion takes place as the strong muscle walls contract to mix and churn the food. The stomach wall produces strong digestive juices and a fairly strong acid which acts upon the food while it is being mixed. Solid food becomes more fluid in the stomach. Digestive *ENZYMES* are proteins that help in breaking down food. These enzymes act chemically to separate food materials into other substances. Special enzymes are found in each part of the digestive system.

The Small Intestine

After it leaves the stomach, the food mass goes to the *SMALL INTESTINE*, the main organ of the digestive system. Over 21 feet long (seven meters), the small intestine does a big job. *The main part of digestion takes place in the small intestine.*

Inside the small intestine are *GLANDS* that help to digest food. Glands are structures that produce and *SECRETE*, or give off, special chemicals. Each of the chemicals does a special job. It acts on only *one kind* of substance. The food that you eat is made mainly of fats, carbohydrates, and proteins. Each part of the food needs to be acted upon and changed by one kind of enzyme or secretion. The *PANCREAS* and the *GALL BLADDER* are glands that get into the digestive act.

The *LIVER* is a very large organ that secretes *BILE*. Bile is stored in the gall bladder, a gland which lies very close to the liver. Bile acts upon fats. The liver stores useful chemicals for short periods. It is a very active organ. In addition to helping with digestion, the liver is very active in energy changes. The liver acts to control the level of a number of substances. The liver changes substances to useful and less harmful forms.

All of the juices from the pancreas, gall bladder, small intestine, and the liver complete the breakdown of food. Finally, the food molecules are small enough and changed enough to move out of the digestive tract through the walls of the small intestines. Capillaries and small blood vessels pick up the digested food molecules. Food materials that have not been digested move on to the *LARGE INTESTINE*.

Inside the Large Intestine

Digestion does not take place in the large intestine. The *LARGE INTESTINE* stores food that is not digested. Water is removed from the food mass and solid wastes are formed. The solid wastes are called *FECES*. Feces are stored temporarily in the last part of the large intestine, the *RECTUM*. Feces leave the body through a body opening called the *ANUS*.

CHECK YOUR UNDERSTANDING

1. What is the first step in the process of digestion?

2. What is a main difference between mechanical and chemical digestion?

3. What are the main digestive organs of the body?

4. What glands and large organs help with the chemical digestion of food while the food mass is in the small intestine?

5. What happens in the large intestine?

> **Key Idea #2:**
>
> - The circulatory system, the respiratory system, and the excretory system transport food molecules, oxygen, and carbon dioxide.

BODY SYSTEMS AND LIFE ACTIVITIES

The respiratory system gets oxygen to the blood stream from the lungs and then, by way of circulation, to body cells. Oxygen in the cells acts on the food molecules. Energy is released from the food. Carbon dioxide gas and water are formed as waste products. The excretory system passes waste products out of the body. Carbon dioxide is released through the lungs.

The circulatory system serves as the transport system for food molecules, oxygen, and carbon dioxide. All of the body systems work hand in hand to get the main jobs of living done. The circulatory system also carries the secretions of the body's glands to the special places where they are needed. The end result of all of the systems working together is that the basic life activities get carried out.

Circulation and the Heart

The *CIRCULATORY SYSTEM* in the human body consists of a pump that forces blood all around the body and a network of tubes that work to move blood. The pump is the *HEART*, and the tubes are the *BLOOD VESSELS*.

The Heart. The *HEART* is the most powerful organ in the body. It is made mostly of muscle tissue. The heart contracts and relaxes in a regular rhythm. The rhythmic beat is known as the *HEARTBEAT*. The beat of the heart is *automatic*. The beat is under the control of a separate system of tissues that regulates the rate of the beat. The heart beats about 70 times a minute when at rest. It beats much faster — even up to 180 times per minute — when you run, jump, swim, or do aerobic exercises.

PULSE is a bulge in the artery walls caused by increased pressure inside the walls as the heart beats. The force of blood in the arteries can be felt as a pulse at the wrist. By holding two fingers on your wristbone below your thumb, you can count the number of times your heart beats in a minute.

How the Heart Works. The heart is about the size of a human fist. It is located between the lungs in the chest cavity. The heart has two sides — a left and a right side. Each side has an upper chamber and a lower chamber. There are four chambers in the human heart.

Parts of the Human Heart

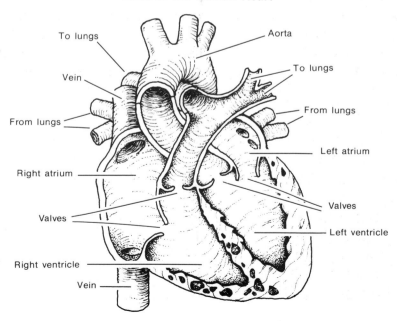

The *RIGHT ATRIUM* and then the *RIGHT VENTRICLE* receive oxygen-poor blood from the body. The right ventricle is muscular and sends the blood to the lungs.

In the lungs the blood gets filled with oxygen. From the lungs, oxygen-rich blood goes to the *LEFT ATRIUM* and then into the *LEFT VENTRICLE.* The left ventricle is a thick muscle and powerful enough to send the blood surging out through the largest of arteries, the *AORTA*, to the entire body.

A stethoscope reveals the sound of the heartbeats as LUPP-DUPP, LUPP-DUPP, LUPP-DUPP, LUPP-DUPP. Hour after hour, day after day, and year after year — the heart pumps its valuable cargo — the life blood of the body.

Blood Plasma. The fluid part of the blood is the *PLASMA.* Proteins in the plasma called *ANTIBODIES* fight diseases. When a foreign substance enters your body, your body reacts by making antibodies. Antibodies attach themselves to the invaders by chemical means. This *DEFENSE SYSTEM* makes an antibody *for each kind* of invading disease. Antibodies give the body *IMMUNITY,* which is the ability to resist disease.

Red Blood Cells and Hemoglobin. Half of the blood is made up of *RED BLOOD CELLS.* Red blood cells are the carriers of oxygen. They have *HEMO-GLOBIN.* Hemoglobin is a protein that joins together *easily* with oxygen. Hemoglobin gives blood its red color. It is an unusual substance because it attracts and holds onto oxygen in the bloodstream. Hemoglobin also gives up oxygen easily for the cells to use. A normal, healthy body has a regular red blood cell count of 4.5 to 5.5 million per cubic millimeter.

Red blood cells

White Blood Cells to the Rescue. *WHITE BLOOD CELLS* make up a much smaller part of the blood. There are normally between 7,000 and 10,000 white blood cells per cubic millimeter. Each white blood cell is much larger than a single red blood cell. The white cells are the body's protection against infection. Some wrap themselves around bacteria. Some make chemicals which kill bacteria. The white cells are able to move out from the bloodstream to where there is an infection. The white blood cell count increases when a bacteria or virus enters the body.

White blood cells

Blood Vessels and Circulation. *BLOOD VESSELS* are hollow tubes that transport blood to and from all parts of the body. The large vessels are like major highways where blood is speeding along at a fast rate. These vessels are *ARTERIES* and *VEINS*. The arteries take blood full of oxygen and food to the smallest blood vessels. Arteries become smaller and smaller as they get farther from the heart. Very small arteries branch out into blood vessels called *CAPILLARIES*. The capillaries are so narrow that they are like small paths in the woods compared to the major highways of the arteries. The walls of the capillaries are one cell layer thin, and the exchange of oxygen and food molecules takes place through the walls. Your body has so many capillaries that all of the millions of cells are each next to a capillary wall. Waste products from cell activity get into the capillaries and then into the smallest of veins on their return to the heart and the lungs. Large veins rush the blood along as it returns to the heart for more oxygen.

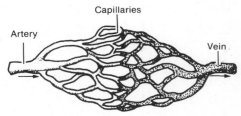

Blood Vessels

Blood Pressure. *BLOOD PRESSURE* is a measurement of how hard your blood pushes against the inside of artery walls. When your heart beats, blood is pushed into the arteries. This forces the artery walls to bulge for a moment. As the large arteries become smaller and smaller, and as blood gets farther away from the heart, more force is needed to keep blood flowing. This makes the heart work very hard. If blood pressure goes up too high, the heart can be damaged.

Platelets and Clotting. Blood *PLATELETS* are the colorless substances in plasma that are helpful in clotting blood. Their action helps form a network of fibers in the blood plasma when there is a cut. Red cells get caught up in these threads and form a clot. Platelets are unusual because they normally do not get involved in clotting in the bloodstream — only when the rough surfaces of a cut appear. Normally there are about 300,000 platelets per cubic millimeter of blood.

Blood platelets

Blood Types. There are four blood types in the world, called Type A, Type B, Type AB, and Type O. The different types are caused by different proteins in the plasma. You have one blood type or another. Sometimes people who are injured or who have certain illnesses may need more blood. Transfusions are given to replace lost blood. *BLOOD TYPING* must be done to identify the patient's particular blood type. If a wrong type of blood is given during a blood transfusion, clotting takes place. Clotting, or clumping, of red blood cells blocks the tiny capillaries. Then the normal oxygen delivery system breaks down. Without oxygen, cells die. Blood banks with a variety of known blood types are set up in hospitals to provide needed blood.

The Respiratory System at Work

The work of the *RESPIRATORY SYSTEM* is to get oxygen from outside your body into your *LUNGS*. The lungs are full of small blood vessels that are able to capture oxygen. When you breathe, the air, full of oxygen, moves into your lungs.

Lungs are the most important organs in the respiratory system. You have two lungs, one in the right side of your chest, and one in the left side. The lungs serve your body by connecting to the outside air and the inside walls of blood vessels.

When at rest, you usually breathe about 14 times a minute. With each breath, you take in about a half a liter of air. A strong sheet of muscle underneath your lungs separates the lung cavity from the abdominal cavity. It is called the *DIAPHRAGM*. The diaphragm is a muscle that helps you breathe. By contracting or pulling, it helps make air move in. It relaxes and moves up, and you breathe out. Your lungs are made of tissue which can stretch. When you *INHALE*, or breathe in, the lungs fill up with air and act like balloons. When you breathe out, or *EXHALE*, the air in the lungs rushes out, and the lungs become smaller. Inhaling and exhaling air mixed with carbon dioxide is a mechanical process. This process depends on changing pressure inside the body.

Air Inside the Lungs. Air comes into your body through two openings; your nose and mouth. Then air travels into one large breathing tube called the *TRACHEA*. The trachea branches into two tubes called *BRONCHI*. One bronchus goes into each lung. Here each bronchus branches out into many tubes called *BRONCHIAL TUBES*. Inside each lung, these breathing tubes continue to branch and become *BRONCHIOLES*.

Finally, at the end of the tiniest bronchioles, there are many sacs of air which are so small that you would have to look through a microscope to see them. They are thin and full of tiny blood vessels. These microscopic air bags inside your lungs are called *ALVEOLI*.

Oxygen: From Respiration to Circulation. Walls of the tiny alveoli air sacs are very thin, and they are usually only one cell layer thick. The walls are always kept wet. Wrapped around the alveoli, just like tiny nets, are many tiny capillaries. Oxygen that is in the air inside the alveoli moves through the walls of the tiny air sacs, through the walls of the tiny capillaries, and into the blood. Oxygen moves from air inside the alveoli to the blood in the capillaries through a process called *DIFFUSION*.

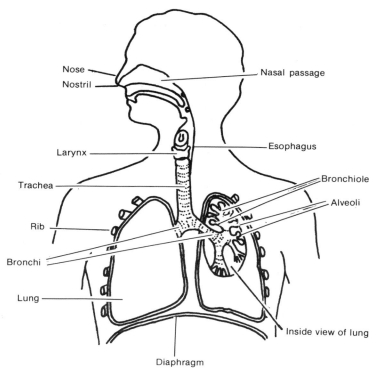

The Respiratory System

The Lungs as an Excretory Organ. The *EXCRETORY SYSTEM* moves wastes from cell activity to the outside of your body. The lungs are excretory organs. When the blood inside of capillaries picks up oxygen in the lungs, it also lets carbon dioxide go into the air inside the lungs. This carbon dioxide is what is left after cells in your body use up food molecules. Oxygen acts upon the food in order to release energy for the basic life activities. Blood full of carbon dioxide returns to the capillaries in the lungs. Then the carbon dioxide is on its way out of the body.

Kidney Structure and Function. The kidneys serve as excretory organs which remove wastes from the blood. There are two kidneys, and they are located in the small of your back, behind your stomach and below the diaphragm muscle. Kidneys are the most important excretory organ. They help maintain chemical balance in the body fluids. They serve as filters to get rid of extra sugar and other waste products in a substance called URINE.

Urine Formation in the Kidneys. Plasma from the blood inside the capillaries goes into tiny tubes called *NEPHRONS*. Wastes and extra water from the plasma stay inside the nephrons. But *much* water, sugar, and minerals from the plasma go back into the capillaries. The wastes which stay inside the nephrons make up *URINE*. Urine is mostly water, with some nitrogen wastes, salts, sugars, and vitamins. Blood continues to go through the capillaries and into veins, which carry it back to the heart. Urine from the nephrons goes into larger collecting tubes in the kidneys to be excreted.

Urine Excretion. There are two tubes which carry urine from each kidney to the urinary bladder. The *URINARY BLADDER* is a muscular bag where urine is stored until it is removed from your body. The bladder stretches as it fills, and then it squeezes together to push urine into a tube called the urethra. The *URETHRA* is the tube which takes the urine to the outside of your body.

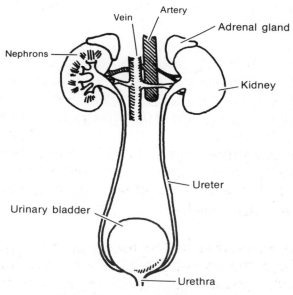

Excretory System (Urinary)

Perspiration Is Formed by Sweat Glands. Your skin is the largest organ in your body. One of the important things the skin does is to get rid of extra water, along with some inorganic salts and urea. This waste material is *PERSPIRATION*. There are thousands of tiny sweat glands in your skin. *SWEAT GLANDS* are coiled tubes which open up at the outside of your skin. These openings are called *PORES*. Perspiration from the sweat glands leaves your body through the pores.

Body Systems Working Together

You have looked at three large systems of the body. It is convenient to study them one at a time, but they work closely together. Circulation, respiration, and excretion are going on at the same time. These three systems all deal with the same substances — blood, food, and carbon dioxide. They are all working together to carry out the basic life activities.

CHECK YOUR UNDERSTANDING

1. What are the two main parts of the circulatory system?

2. What do red blood cells do to help the body?

3. What do white blood cells do to help the body?

4. What are the three kinds of blood vessels?

5. What is blood pressure?

6. What is the main pair of organs of the respiratory system?

7. What is the diaphragm?

8. What is the process by which oxygen gets into the blood in the capillaries?

9. What is the process of excretion?

10. What are the three excretory organs of your body?

> **Key Idea #3:**
>
> - The nervous system, sense organs, and endocrine glands send and receive information to control and coordinate the body.

COORDINATING SYSTEMS OF THE BODY

The nervous system, sense organs, and endocrine gland system are all coordinating and regulating parts of the body. These systems send messages from one place to another. They tell body parts what to do. They inform the body about what is happening in the outside world. They also regulate body functions because of changes they cause inside the body. They work together to make sure that the body is in balance.

The Nervous System

The human *NERVOUS SYSTEM* is better than the most expensive computer on the market. It is efficient and requires little maintenance, although it runs night and day. This amazing human computer consists of the *CENTRAL NERVOUS SYSTEM* and the *PERIPHERAL NERVOUS SYSTEM*. The central nervous system has two parts: the *BRAIN* and the *SPINAL CORD*. The peripheral nervous system consists of all the nerves that transmit messages between the central nervous system and other parts of the body.

Three Parts of the Brain. The three parts of the brain are the *CEREBRUM*, the *CEREBELLUM*, and the *BRAIN STEM*. The largest part of the brain is the cerebrum. The cerebrum controls the way in which you think, learn, remember, and feel about things. Body movement is also controlled by the cerebrum.

The cerebrum is divided into two halves, or *HEMI-SPHERES*.   The left hemisphere of the cerebrum controls activities on the right side of the body. The right half of the cerebrum controls activity of the left side of the body. In addition, the left side of the brain has more to do with math, language, and logical thinking. The right side is connected with the musical and artistic part of the world.

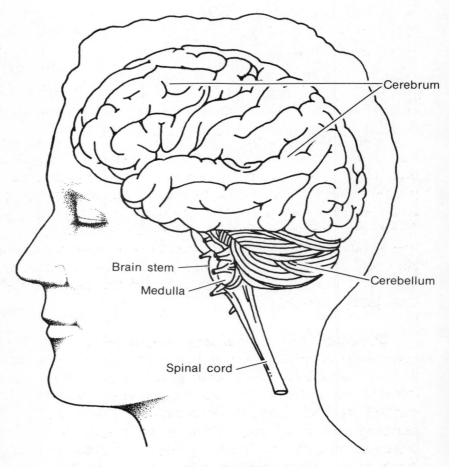

Parts of the Brain

The cerebellum is the smaller part of the brain which is underneath the cerebrum. The cerebellum controls balance and helps coordinate muscular activity such as walking and writing. The cerebellum controls these activities by sending and receiving messages between the sense organs and the muscles.

Under the cerebellum is the brain stem, which connects the brain and the spinal cord. A part of the brain stem called the MEDULLA controls the automatic activities of your body such as heart rate, gland secretions, digestion, respiration, and circulation. You do not think with your medulla. It is always on "automatic." Otherwise you would spend all of your time operating the parts of your body.

The Spinal Cord. The SPINAL CORD is a thick bunch of nerves starting at the medulla and going down the back, inside a bony spine. The spinal cord is protected by the bony vertebrae. The brain sends out and receives information through the spinal cord. Thirty-one pairs of SPINAL NERVES branch off from the spinal cord. The spinal nerves send nerve impulses all over your body. The spinal cord, brain stem, and the brain are the central controls of the sense organs and body systems.

Sending and Receiving Messages. Your body constantly receives information from the outside world through your eyes, ears, and other sense organs. That information goes to the central nervous system. The central nervous system responds to these signals by sending out impulses. Your body responds to these impulses in different ways. It reacts in order to protect itself, to maintain itself, and to continue to do all of the basic life activities. Some reactions are automatic. Other reactions that require thought take more time.

What Neurons Do. Nerve cells are called *NEURONS*. Information travels from one neuron to the next by an *IMPULSE*. An impulse is a very small electrical charge which moves very, very fast. There is a short gap between neurons called a *SYNAPSE*. Within each neuron an impulse travels from end to end. When an impulse comes to the end of one neuron, there is a chemical change at the synapse: the next neuron picks up the impulse. Information moves from one place in your body to the next by traveling along lengthy sets of neurons.

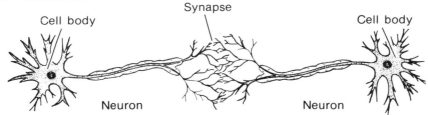

Cell body   Synapse   Cell body

Neuron   Neuron

Three Kinds of Neurons. There are three kinds of neurons in your nervous system; *SENSORY NEURONS*, *ASSOCIATION NEURONS*, and *MOTOR NEURONS*. Sensory neurons send impulses from sense organs to the spinal cord or the brain. Association neurons in the brain and spinal cord carry the impulses from sensory neurons to motor neurons. Motor neurons carry impulses from the brain and spinal cord to muscles and glands.

Reflex Actions. Sneezing, coughing, and blinking your eyes are *REFLEX ACTIONS*. They happen automatically. What happens if you accidentally touch a frying pan which is very hot? *SENSORY NEURONS* pick up the "It's hot!" message from the heat and the feeling it causes. The neurons send impulses to the nerve cord or *SPINAL CORD*. Inside of the spinal cord, *ASSOCIATION NEURONS* read and interpret the message and send impulses out to the *MOTOR NEURONS*.

All of this happens in an instant. *You pull your hand away very quickly.* You have been saved from a serious burn by the reflex actions of your nervous system and muscle system.

In a similar way, if some object comes flying toward your eyes, you blink, blink, blink — without thinking! Reflex actions are *AUTOMATIC*, and they are *fast.* They are mostly built-in protective actions.

Sense Organs

The body connects with the outside world through special organs called *SENSE ORGANS.* The five main sense organs are the: 1) eyes, 2) ears, 3) skin, 4) nose, and 5) tongue and mouth.

Eyes and Light. The *EYES* of animals are their cameras to the outside. Highly specialized, they focus light in all kinds of conditions. Light enters the eye through the *CORNEA*, which is a clear, transparent covering layer that is surrounded by the *SCLERA*, or "the white of the eye."

The color of your eyes is in the *IRIS.* The Greek goddess of the rainbow was named Iris. The iris contracts the *PUPIL*, or opening of the eye. This controls the amount of light that enters. The iris has muscles to control the size of the pupil opening. Your pupils get larger when you enter a dark room. The larger opening lets in more light. More light on the subject makes you see things more clearly. Likewise when you go into the bright sun of a sunny day, the pupils shut down to very small openings. This keeps *too much* light from entering. Too much light or glare sends too many signals to the nervous system, and that blots out the image. The bright light of direct sunlight can damage the receiving screen in the back of the eyeball. You should never look directly into the sun.

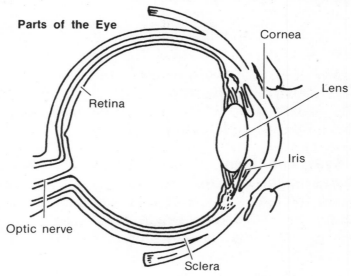

Parts of the Eye

Cornea

Lens

Retina

Iris

Optic nerve

Sclera

The Lens and Retina. There is a clear, soft *LENS* behind the pupil opening in the iris. Muscles on the lens cause it to change its shape. By changing its shape, the *LENS* changes the focus of light rays on the inside wall of the eye.

The *RETINA* is the inside wall of the eye and is covered by special receptor cells and nerve cells. The retina receives information from the light rays which focus upon it. The retina sends this information to sensory neurons. The sensory neurons form the *OPTIC NERVE*, which goes to the brain. The brain gives meaning to the impulses from the optic nerve and forms a picture. All of this happens very quickly. Light travels fast, and so do nerve impulses.

The Sense of Hearing. Your *EARS* pick up and send information about sounds to your brain in the form of nerve impulses. Sounds are collected in the outer ear and are sent into the ear canal to the eardrum. The *EARDRUM* is a thin tissue which separates the outer ear from the middle part of your ear. Three small bones in the middle part of the ear make sounds louder.

In the *inner* part of the ear there are spaces filled with liquid. One of the liquid-filled spaces of the inner ear is the *COCHLEA*, the *real hearing organ*. The cochlea has hearing receptor cells inside which are like tiny hairs. Sounds from the middle part of your ear cause liquid inside the cochlea to move around. Movement of this liquid bends the hair-like receptor cells. The receptor cells send impulses to the *AUDITORY NERVE* which goes to the brain. The brain receives impulses from the auditory nerve and gives meaning to the sound impulses. Loud, soft, high, low, melodic, or off-key, your ears are really "fine-tuned."

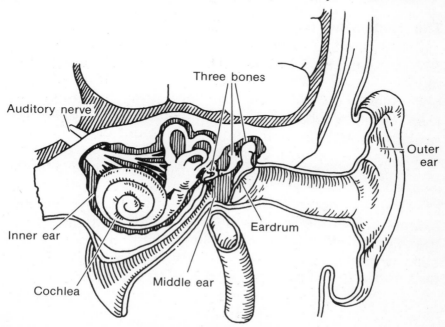

Parts of the Ear

Skin as a Sense Organ. The largest organ in the body is the skin. This very widespread organ has *RECEPTORS* that provide different kinds of information. They send messages for heat, cold, heavy touch, light touch, pressures, and pains. From the tip of your toes to the top of your head, your skin is able to pick up messages and send them to the nervous system. There are billions of receptor cells in the layers of skin that cover the body. The skin is the body's first line of defense against disease. It is also the first line of defense in warning of any kind of danger, such as hot frying pans, sharp edges, or harsh chemicals.

Taste and Smell. The senses of taste and smell are alike. They both have receptor cells which react to chemicals in food. *TASTE BUDS* on the tongue pick up four basic kinds of taste: sweet, sour, bitter, and salty.

The sense of smell is more exact. Receptors for smell react to thousands of different chemicals.

The Endocrine Glands

A system of glands known as the *ENDOCRINE SYSTEM* releases *chemical messengers* directly into the blood stream. The chemical messengers are called hormones. *HORMONES* are very powerful, and they help regulate what happens in tissues and organs all over the body. Growth and development of body parts is under the control of hormones. Levels of different chemicals are regulated by hormones.

The hormone *INSULIN* from the pancreas is an example. Insulin is in charge of the level of sugar in the blood. Without insulin, sugar cannot get into the cells where it is needed. The sugar level in the blood gets too high. Diabetes, a destructive disease, results.

Too much insulin "burns up" the sugar in the blood, and cells cannot obtain the energy they need to operate. The brain is one of the first organs affected.

In general, hormones speed things up or slow things down. Keeping an internal balance is an important role played by hormones from the endocrine system. The body does not store hormones. More than twenty different hormones have an effect on everything from the development of sex characteristics, to kidney function and regulation of stress reactions, such as heart rate and blood pressure.

The basic life activities are regulated in many ways by secretions from the endocrine glands. It is very important to have the right hormones at the right times in order for your body to carry on basic life activities.

CHECK YOUR UNDERSTANDING

1. What are the three functions of your nervous system?
2. What are the two parts of your central nervous system?
3. What are the three parts of your brain?
4. How does information travel inside your nervous system?
5. What are the five sense organs of your body?
6. What is the retina of your eye?
7. What are five different kinds of messages which receptor cells in your skin can send to your brain?
8. What are the four basic kinds of tastes?
9. Where are receptors for taste located?
10. Which part of the ear contains receptor cells for hearing?

> **Key Idea #4:**
>
> • The skeletal and muscular systems provide support, shape, and protection. They work together to make body movements.

THE HUMAN SKELETON

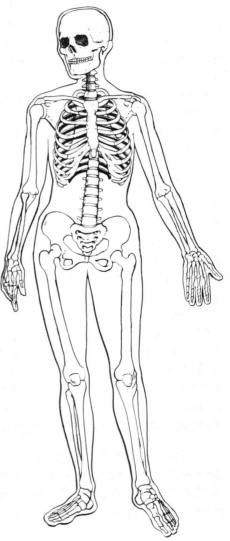

The 206 bones of the human body are strong and still light enough for ease of movement. They form the *SKELETAL SYSTEM.* The bones give support to the body and provide it with a shape. The mixture of big bones, small bones, flat bones, wide bones, and specially shaped bones of the spinal cord give the body ways to do many different things.

The bones form a framework that surrounds the inner organs and gives them protection. The vital lungs and heart are encased in a rib cage. This keeps them out of danger.

The bones are circular and hollow. This structure is one of nature's strongest forms. Inside of the bones there is a substance called MARROW. It serves, in some bones, as the source of red and white blood cells. Chemicals such as phosphates and calcium are essential to certain activities and are stored and released by the bones as needed.

Cartilage to Bone. Most bones start out as cartilage. CARTILAGE is a tough tissue which is not as hard as bone. As your body grows, more and more cartilage is changed into bone. This process goes on all your life, but most changes happen before you are twenty-five years old. The process of replacing cartilage cells with bone cells is called OSSIFICATION. Not all cartilage changes to bone. Feel the end of your nose. It is still cartilage and will never ossify.

Joints. JOINTS are the places where bones are connected to one another. Cartilage covers bones at the joints. This cartilage acts like a cushion and protects bones from rubbing against each other. At the movable joints the bones are connected to one another by special tissues called LIGA-MENTS. Ligaments stretch in order to allow the bones to move. At the same time, ligaments hold the bones together.

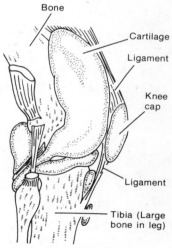

The Knee Joint

Two important kinds of joints for movement in the skeletal system are the HINGE JOINT and the BALL AND SOCKET JOINT. The joint in your elbow is a hinge joint. The joint in your shoulder is a ball and socket joint. Some joints can only move a little, like your rib and spine joints. Some joints, such as those in your knees, move a great deal.

The Muscle System

Muscles work by contracting or tightening up and then relaxing or resting. Muscles are for movement. Movement of the arms, legs, and torso are the big movements. Other muscle movements take place in the heart, the digestive system, and in the rib cage and diaphragm.

Most bones have muscles attached to them. These muscles act with the bones. The bones serve as levers to exert forces started in the muscles. This is a good example of how *all of the systems of the body work together.*

Muscles are either *VOLUNTARY* or *INVOLUNTARY.* Skeletal muscles, like those in the arms and legs, are voluntary, because you *can make* them work or rest. You can wiggle your toes or wave your hands. You *cannot* control involuntary muscle, such as the muscles of your heart. Involuntary muscles work automatically.

Skeletal Muscles Are Voluntary. *SKELETAL MUS- CLES* are voluntary muscles which are attached to bones or to one another. You work these muscles to move your bones, and make your body go places and do things. Skeletal muscles bend and extend. Skeletal muscles cause movement only by pulling on bones; never do they push bones.

The large muscles of the upper arm of humans is a good example of muscles attached to bones and the pulling action that makes the arms move. The *BICEPS MUSCLE* is the one in the upper arm that bulges to the front. When showing off muscles, this is the one most often flexed.

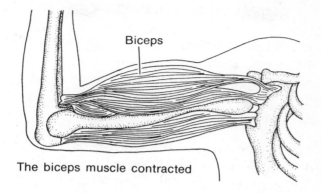

Biceps

The biceps muscle contracted

Involuntary Muscle. *CARDIAC MUSCLE* is involuntary muscle that is only in the heart. It contracts and relaxes automatically.

Another kind of muscle which contracts and relaxes automatically is *SMOOTH MUSCLE*. Smooth muscle is involuntary muscle that is found in the walls of digestive organs and blood vessels. You do not need to call on involuntary muscle to do its job. It works on its own.

CHECK YOUR UNDERSTANDING

1. What are four functions of the skeletal system?
2. What substance in the bone makes new blood cells?
3. What is ossification?
4. Name a hinge joint.
5. Name a ball and socket joint.
6. How do muscles make things move?
7. What are the three kinds of muscle in your body?
8. Where is cardiac muscle found?
9. What kind of muscle is located in the walls of the digestive organs and blood vessels?

SUMMARY OF CHAPTER 8

The digestive system makes food ready to be used by the cells. Food that is ingested must be made small enough and simple enough to be used by the cells in your body. Digestion makes food into usable food molecules.

The heart pumps blood through vessels to cells in your body. Blood carries food molecules, oxygen, special kinds of blood cells, and chemicals. Oxygen enters your lungs with the air that you breathe. In the lungs, oxygen diffuses into the blood. Carbon dioxide from the cells is breathed out. Your blood, lungs, kidneys, and skin remove wastes from the cells to the outside of your body.

Sense organs send information to the central nervous system. Nerve impulses travel from one neuron to another as tiny electrical charges. Your central nervous system gives meaning to this electric information and sends out impulses on nerves that cause you to react.

Your endocrine glands make chemical juices which they put into your blood. These chemicals regulate and balance the way that parts of your body work.

The muscles and bones in your body support your weight. They give shape to your body while protecting softer organs inside. Muscles relax and contract to make your bones work together like levers, so you can walk, jog, run, or do gymnastics.

Each of the systems of the body has a different function to perform. Yet, all body systems work together in order to carry out the basic life activities.

WORDS TO USE

1. ingestion
2. digestion
3. food molecules
4. mechanical digestion
5. chemical digestion
6. salivary glands
7. saliva
8. esophagus
9. stomach
10. enzyme
11. small intestine
12. gland
13. secrete
14. pancreas
15. gall bladder
16. liver
17. bile
18. large intestine
19. rectum
20. anus
21. circulatory system
22. heart
23. blood vessels
24. heartbeat
25. pulse
26. right atrium
27. left atrium
28. right ventricle
29. left ventricle
30. aorta
31. arteries
32. veins
33. capillaries
34. blood pressure
35. blood platelets
36. blood type
37. blood typing
38. plasma
39. antibodies
40. defense system
41. immunity
42. red blood cells
43. hemoglobin
44. white blood cells
45. respiratory system
46. lungs
47. diaphragm
48. inhale
49. exhale
50. trachea
51. bronchi
52. bronchial tubes
53. bronchioles
54. alveoli
55. diffusion
56. kidneys
57. skin
58. urine
59. perspiration
60. excretory system
61. urethras
62. urinary bladder

WORDS TO USE

63. sweat glands
64. pores
65. nervous system
66. central nervous system
67. peripheral nervous system
68. brain
69. spinal cord
70. cerebrum
71. cerebellum
72. brain stem
73. hemisphere
74. medulla
75. spinal nerves
76. neurons
77. impulse
78. synapse
79. sensory neurons
80. motor neurons
81. association neurons
82. reflex action
83. sense organs
84. eyes
85. cornea
86. sclera
87. iris
88. pupil
89. lens
90. retina
91. optic nerve
92. receptors
93. taste
94. taste buds
95. smell
96. ears
97. eardrum
98. cochlea
99. auditory nerve
100. endocrine system
101. hormones
102. insulin
103. skeletal system
104. marrow
105. cartilage
106. ossification
107. joints
108. ligaments
109. hinge joint
110. ball and socket joint
111. muscle system
112. voluntary muscle
113. involuntary muscle
114. skeletal muscles
115. bicep muscle
116. cardiac muscle
117. smooth muscle

• REVIEW QUESTIONS FOR CHAPTER 8 •

1. What is the difference between ingestion and digestion?

2. What is the final thing that food gets to be during digestion?

3. What is the difference between mechanical and chemical digestion?

4. What do the salivary glands do?

5. What happens to food in the stomach?

6. What is special about the small intestine?

7. What makes the liver a special organ?

8. What happens in the large intestine?

9. What is the main organ in the circulatory system?

10. What is hemoglobin? What does it do?

11. What do the white blood cells do?

12. What do the blood platelets do?

13. Why is it important to know a patient's blood type?

14. How are capillaries different from arteries?

15. What happens in the capillaries?

16. What is meant by blood pressure?

17. What is a pulse?

18. What is the main work of the respiratory system?

19. What part does the diaphragm play in the respiratory system?

20. What is the difference between the trachea and the bronchi?

21. What do the kidneys do?

• REVIEW QUESTIONS FOR CHAPTER 8 •

22. What does the urinary bladder do?

23. How does the skin serve as an excretory organ?

24. What is the job of the nervous system, sense organs, and the endocrine system?

25. What is another name for nerve cells?

26. What is the main difference between a sensory neuron and a motor neuron?

27. What is the best way to describe a reflex action?

28. What is the job of the sense organs?

29. How does the pupil regulate the amount of light that enters the eye?

30. What does the brain do after it gets impulses from the optic nerve?

31. What is the function of the receptors in the skin?

32. What is the cochlea? What do its receptor cells do?

33. What are hormones? What do they do?

34. What does the skeletal system do for the body?

35. What is ossification?

36. How do muscles work?

37. What is the difference between voluntary and involuntary muscles?

Chapter 9

Describing Patterns In Reproduction, Growth, and Development

Chapter Goals:

To understand how organisms produce offspring like themselves.

To recognize that there are patterns in the processes of reproduction, growth, and development of new organisms.

Key Ideas:

- The offspring of each animal species are like the parents, but are not always identical.

- Sexual reproduction is a duplicating process that transmits genetic material to the next generation.

- Organisms go through stages of growth and follow patterns in their development.

- Reproduction, growth, and development in humans follow a pattern much like that of other mammals.

INTRODUCTION

Every species must reproduce in order to maintain a population. Individuals eventually die; but, because of reproduction, the population lives on. Parent organisms create offspring like themselves; but, in sexual reproduction, the offspring are not exactly like either parent. The offspring get half of their genes from each parent. This method of reproduction creates individuals that are truly unique. Humans and other mammals use sexual reproduction.

Key Idea #1:

- The offspring of each animal species are like the parents, but not always identical.

LIFE FROM OTHER LIFE

What would happen to a species that did not reproduce? Sooner or later, every last member would die, and the species would be gone — *EXTINCT*. All living species keep themselves going by means of reproduction. In fact, the ability to reproduce is one of the basic life activities for a species.

Spontaneous Generation

It may seem obvious that rabbits produce more rabbits and giraffes produce more giraffes; but, many years ago, some people believed in *SPONTANEOUS GENERATION*. They saw flies come from rotten meat and frogs come out of the mud of a pond in the spring, so they reasoned that rotten meat makes flies, and mud makes frogs. Early scientists disproved that idea. It was discovered that eggs laid in meat and on the edges of ponds were the source of flies and frogs. Spontaneous generation does not exist. Life comes only from other life.

DNA Stores the Patterns of Life

Each animal's cells contain the information needed to make another animal just like itself. Each and every cell contains *CHROMOSOMES* located in the cell *NUCLEUS*, or center. These chromosomes are made of *DNA*, a most important life chemical. Without DNA, there would be no life at all.

DNA stores information in a pattern, much as a book stores information in a pattern of letters and words. The pattern in a cell can be read by other chemicals to make a new cell. The patterns in the sex cells of rabbits contain all the information needed to make rabbits after their own kind. Cottontail rabbits and jack rabbits are similar, but their DNA makes them into cottontails or jack rabbits.

Most Reproduction Requires Two Parents

One rabbit cannot reproduce by itself. Some very simple animals can, but all of the more complicated species require two parents — a male and a female. Have you ever heard the expression, "He's a chip off the old block," referring to a child who resembles a parent in some particular way? Have you ever seen a child who was an exact copy of one of the parents? Because you all have two parents and get exactly half of your DNA from each, you are never an exact duplicate of one parent. You may resemble one more closely than the other, but you always carry traits from both parents. Humans are really chips off two old blocks, the mother and the father.

The Advantage of Diversity

It is much safer for the survival of a species if the population is made up of individuals with a range of differences than for them to be all identical. Variety or *DIVERSITY* gives the population better chances to *ADAPT*. In case of a disaster, there is a much greater chance that some individuals will have traits that allow them to meet with changes in the environment. These individuals will survive to carry on the species. Some individuals in the population will not be able to adapt to sudden environmental changes. Such individuals will die and will not be able to reproduce.

Adaptation and Resistance

Suppose a disease sweeps through a population of deer, killing many members. If none of them is able to fight the disease, then all of the deer would die. Chances are that a few, at least, will be *RESISTANT* to the disease, will fight against it, and will survive. You can say that resistance is a kind of adaptation. Adaptations come from different combinations of genes and DNA.

Insects have vast numbers of offspring, each slightly different from the next. Often they become a problem in farm areas by eating crop plants. Spraying chemicals never seems to kill all of the pest insects. No matter what spray is used, some seem to be resistant to its effects. Of course, those are the only ones who live to reproduce. The next season, those insects with resistance to the spray reproduce offspring like themselves. More of the population is resistant to the chemicals. After a while the chemical spray that was effective becomes completely ineffective. A new chemical with different powers must be developed. All of the insects became resistant to the first chemical.

CHECK YOUR UNDERSTANDING

1. What is the purpose of reproduction?

2. What is spontaneous generation?

3. What is DNA, and what does it do?

4. What advantage is there to having variations among individuals in a population?

5. What causes you to resemble your parents and grandparents more than you resemble other people?

Key Idea #2:

- Sexual reproduction is a duplicating process that transmits genetic material to the next generation.

SEXUAL REPRODUCTION

Sexual reproduction is the type of reproduction that most animals use. Sexual reproduction involves the duplication of the chromosomes of the cell and the division of the cell itself. Although cells copy themselves and divide all the time, in sexual reproduction the process is not the same.

Cells Divide to Make New Cells

The basic structure of living things is the cell. A common feature of cells is that they divide. Skin cells, bone cells, muscle cells, blood cells, and the lining cells of body organs must all reproduce themselves. The main actor in the process is the nucleus. In the nucleus of each cell is a set of chromosomes. The number of chromosomes is a set number for each species. Fruit flies have four pairs of chromosomes, and humans have 23 pairs.

Chromosome Duplication

The first step in the making of a new cell is the copying or duplicating of the old cell's DNA in its chromosomes. Each new cell will need a complete copy. Division into two new cells can proceed only when gene materials are in the right amounts. For a short time, just before cell division occurs, a cell has two complete sets of chromosomes.

Mitosis

After the chromosomes have been duplicated, they line up at the center of the cell. Soon each double pair of chromosomes separates, one pair going to one end of the cell and the other going the opposite way. Then the cell membrane pinches in between two sets of chromosomes, separating the original cell into two identical "daughter" cells. The process of *MITOSIS* is complete. Each daughter cell has the species number of chromosomes.

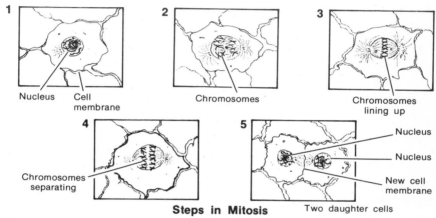

1

Nucleus   Cell
          membrane

2

Chromosomes

3

Chromosomes
lining up

4

Chromosomes
separating

5

Nucleus

Nucleus

New cell
membrane

Steps in Mitosis   Two daughter cells

Asexual Reproduction

If the organism in question consists of only a single cell, mitosis reproduces the entire organism. This process is called *ASEXUAL REPRODUCTION*, meaning reproduction without sex, because only one parent is involved.

Asexual reproduction has one advantage — speed. A bacterium or other single-cell protist can reproduce every twenty minutes or half-hour. The chief disadvantage of asexual reproduction is its lack of diversity. All the offspring are exact copies who will respond to changes in the environment in the same way. If there is a dangerous chemical in the environment that kills one of the organisms, it will probably get them all.

Sexual Reproduction in the Animal Kingdom

The majority of animal species in the world use *SEXUAL REPRODUCTION*. Sexual reproduction involves the genetic materials from two individuals. This guarantees more diversity and variety of traits and characteristics than in asexually produced organisms. All of the vertebrates and all of the insects use sexual reproduction. These groups are the most successful organisms in the world.

Gametes from Gonads

All higher animals reproduce sexually using two parents of different sexes, male and female. The female produces *EGG CELLS* from an organ called an *OVARY*. The male produces *SPERM CELLS* from organs called the *TESTES*. The sex organs, the ovary and testes, are called *GONADS*. The sex cells that the gonads produce, the egg and the sperm, are called *GAMETES*.

Unlike a chicken egg, an *egg cell* is too small to be seen without a microscope. The hen's egg does contain an egg cell, a fertilized one, but most of the egg is food for the developing chick.

Male animals produce millions of sperm cells. Egg cells are rather large as cells go, but sperm are very small cells, about one thousandth the size of an egg cell. Eggs contain food for the early stage of the developing offspring. Sperm are mostly a nucleus, with a half set of chromosomes.

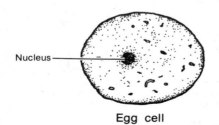

Nucleus

Egg cell

Sperm cell

Meiosis

Both egg cells and sperm cells are single cells containing only half as many chromosomes as the parents they come from. Neither can live by itself for more than a few days. The gametes are formed by a process called *MEIOSIS* (my oh sis). Meiosis is similar to mitosis, but with more duplicating and dividing. Producing eggs and sperms is more complicated.

As in mitosis, the first step in meiosis is the copying and duplicating of the DNA. The cell divides in the usual way. Then the cell divides a second time *before* duplicating the DNA. Four daughter cells are formed, each with half the usual amount of chromosomes. This set of daughter cells becomes the egg cells and the sperm cells. Meiosis is used *only to produce gametes*, so it occurs only in the ovaries and testes.

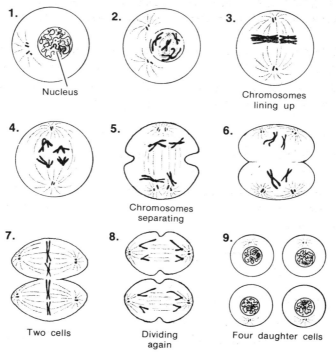

1.

Nucleus

2.

3.

Chromosomes lining up

4.

5.

Chromosomes separating

6.

7.

Two cells

8.

Dividing again

9.

Four daughter cells

Steps in Meiosis

Fertilization

Like the egg, each sperm has half the usual number of chromosomes and cannot survive on its own. Unlike the rather large, round, and slow-moving or stationary egg, the sperm propels itself forward by means of a tiny tail which it rapidly whips back and forth. Sperms are very small compared to egg cells. Many species produce millions of sperms at one time. Although all the sperms move toward an egg cell, only one sperm can *FERTILIZE* the egg. One sperm gets through the cell wall of the egg to join with the egg, fertilize it, and create a new cell. The nucleus of the egg cell and the nucleus of the sperm cell each provide half the chromosomes. Together, the egg and the sperm provide the newly formed cell with a full number of chromosomes. A full number of chromosomes is essential to making a new individual.

The Zygote Is a Fertilized Egg

Imagine an egg of a fish attached to a rock on the bottom of a stream. A female fish has laid her eggs. Each egg is a tiny speck, but it carries the possibility of a new fish within it. It is surrounded by hundreds of other eggs just like it. Above the mass of eggs swims a male fish. He releases sperm cells into the water, billions of them, and they begin a frantic race against time. Each swims through the water toward an egg. One sperm reaches an egg just ahead of dozens of competing sperm. The successful sperm drives itself right through the outer *MEMBRANE* of the egg, piercing this layer. (All animal cells have a containing structure called a membrane.) The membrane quickly seals the puncture. Almost immediately, the rest of the membrane becomes so tough that no other sperm can come through. Unused sperm and unfertilized eggs do not survive.

Once inside the egg, the sperm continues to move and meets the nucleus and the chromosomes of the egg cell. The sperm's own membrane breaks open, and the chromosomes from the nucleus of each of the two cells unite. The egg is now fertilized. The fertilized egg, called a ZYGOTE, begins to develop into a new fish.

External and Internal Fertilization

The process of fertilization described above is called EXTERNAL FERTILIZATION because the egg is fertilized outside the body of the female. Most fish and amphibians use external fertilization, but some fish and all reptiles, birds, and mammals use INTERNAL FERTILIZATION, in which the male puts sperm inside the female's body, where fertilization occurs.

CHECK YOUR UNDERSTANDING

1. What is the first step in mitosis?
2. What is a daughter cell?
3. What is the main disadvantage of asexual reproduction?
4. What are the two methods by which hydras reproduce?
5. What type of reproduction do most animals use?
6. What is the difference between egg sex cells and sperm sex cells?
7. What is the difference between mitosis and meiosis?
8. What happens after the sperm gets through the cell membrane of an egg?
9. What is a zygote?
10. What is the difference between internal and external fertilization?

> **Key Idea #3:**
>
> - Organisms go through stages of growth and follow patterns in their development.

GROWTH AND DEVELOPMENT OF ANIMALS

An organism begins as a zygote and goes through stages of development until it becomes an adult. Cells duplicate and divide to form new cells. The organism grows. If it is a mammal, when it reaches a certain developmental stage, it is ready to live outside of the mother's body.

From Zygote to Embryo

A zygote is still only a single cell, but it has a full set of chromosomes. Many things must happen before the zygote becomes an adult. In the case of a fish, the zygote divides to form two identical but smaller cells attached to each other. This process is repeated many times, forming four, then eight, then sixteen cells. In a few days, there are millions of different cells. The zygote grows into an *EMBRYO.* An embryo is an early stage in the development of an organism.

Fertilized egg   Two cells   Four cells   Eight cells   Many cells

Development

Gradually, the new cells that are formed take on different shapes and functions. This is called *DEVELOP-MENT.* Cells at one end of the developing embryo begin to form part of the head — eyes, mouth, gills, and other organs. Those on one side become nerve tissue that eventually make up the spinal cord. Inside, the heart and other organs develop. Signals in the chemical DNA are causing differences in cells. This is *CELL DIFFERENTIATION.* Eventually, a complete fish is formed. It is tiny (no bigger than the original egg), but cell differentiation has resulted in all the parts necessary for a fish. Not just any fish, but a fish of the same species as the parents.

Growth

This young fish, still curled inside its egg, has a difficult job ahead. It is so small that it has little chance of survival. Other organisms in the environment compete for food, and developing eggs are a good target. It must grow rapidly or die.

The fish hatches and begins to eat. Anything small enough to go in its mouth is fair game — algae, baby shrimp, any kind of microscopic water creature it can catch. It, too, is fair game for larger fish and other animals. Many of its brothers and sisters become meals for turtles, birds, and fish. As it eats and grows, its chances for survival improve. If it escapes being food for another organism, it will live long enough to be able to be a part of the reproduction of the species. Most of the hatch are not so lucky. That is why fish must lay hundreds and thousands of eggs in order for a few to survive. Successful reproduction, development, and growth are essential for a species to survive.

Development in Mammals

Mammals are very different from fish, but their development starts out the same way. A big difference is that mammals take better care of their offspring for longer periods of time. Protection from danger and preparation for living by themselves give mammal offspring a better chance than fish. Therefore, fewer individuals need to be born to ensure survival of the species.

Like all other animals, mammals produce eggs and sperms. *The vast majority of mammals do not lay eggs.* The egg cells develop and are fertilized inside the body of the mother. There are only two egg-laying mammals; the spiny anteater, and the duck-billed platypus. Both animals live in Australia.

Stages of Development

Some offspring look like their parents right from the start. A newborn calf, a young colt, or a baby giraffe are not mistaken for some other animal. The organisms that develop from frog's eggs, however, are tadpoles and do not look like frogs at all. Bullfrogs are the same. They go through stages where they are not like the parents.

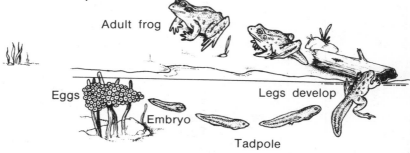

Adult frog

Eggs   Legs develop

Embryo   Tadpole

Butterflies, moths, and beetles go through stages where they are not at all like insects. Butterflies and moths have a caterpillar stage and a cocoon stage. Beetles develop from the pupa stage into adult beetles.

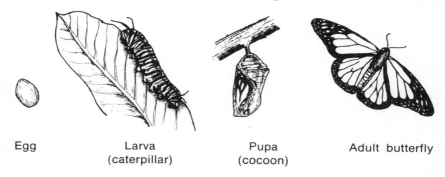

Egg   Larva (caterpillar)   Pupa (cocoon)   Adult butterfly

Grasshoppers and praying mantises hatch as small grasshoppers and mantises. They just get bigger and bigger, but look like their parents right from the start. Babies of human beings look like miniature humans for the most part. They just grow bigger, become children, go through some major changes during their teens, and reach maturity as young adults.

Larger Offspring

Mammal babies may seem small and cuddly to us, but they are gigantic compared to a baby fish, which is smaller than a fingernail. The fish must depend on the small food supply in its egg, but the mammal embryo gets food from its mother throughout its development. For this reason, the mammal can grow much larger and develop more fully before facing the outside world.

The Mammal Uterus

Every mammal female, from cats and dogs to horses, whales, and people, has a *UTERUS* inside her body. The uterus is an organ designed to hold, feed, and protect a developing embryo, and to push it out when its development is complete. The egg is deposited in the uterus and is fertilized there by sperm from a male. As the zygote develops, it gets food from the mother and is also protected because it is inside her body.

Inside the uterus, the zygote forms protective membranes around itself, making a fluid-filled sac for the embryo. Parts of these membranes form a *PLACENTA*. A placenta is a special organ that gets food from the mother's uterus and passes it to the developing embryo. Most mammal species form placentas.

Food for the Young Mammal

A mammal egg is much smaller than a hen's egg or even a fish egg, because it does not need a food supply for more than a few days. The food that it gets from the mother is much more than could ever be packed into an egg.

After birth, the mammal offspring continues to get nourishment from the mother in the form of milk. No other type of animal cares for and feeds its young as mammals do. As a result, most mammals have few offspring because they can expect many of them to survive.

Gestation Differences

All mammal embryos follow the same basic pattern of development. However, the *GESTATION* time required varies greatly depending upon the size of the animal. Gestation time is the time from the fertilization of the egg until birth. Gestation times for a range of mammals is shown in the following chart.

Mammal Gestation Times

Mammal	Approximate Number of Days
Mouse	20
Rabbit	31
Cat or Dog	63
Monkey	210
Human	**275**
Cattle	281
Horse	336
Whale	360
Elephant	624

A whale baby takes almost a year to develop, and an elephant more than a year and eight months. That is a long time for a mother to carry a baby! In general, the larger the mammal, the longer the time required for development.

Some animals are exceptions. For instance, the *opossum*, which is larger than a cat, has a gestation period of only 13 days. That is because an opossum is a *MARSUPIAL* mammal, not one of the more common placental kind. In marsupial animals, the embryo in the uterus gets no food from the mother. Instead it must be "born" at a very early stage. Then, the embryo goes into a pouch on the mother's body and nurses milk for a long time while it completes its development.

CHECK YOUR UNDERSTANDING

1. What is cell differentiation?
2. What is the difference between the way young frogs and young grasshoppers resemble their parents?
3. Why do fish lay hundreds of eggs at one time?
4. What does the uterus do for the developing embryo?
5. What is a placenta?
6. What is meant by gestation time?
7. Why do marsupial animals have short gestation times?

> **Key Idea #4:**
>
> • Reproduction, growth, and development in humans follow a pattern much like that of other mammals.

HUMAN REPRODUCTION, GROWTH, AND DEVELOPMENT

Humans are mammals with the same reproductive organs as other mammals. Humans use sexual reproduction and internal fertilization. Like other mammals, the human embryo develops inside the uterus of the female. After birth, the baby gets nourishment from the *MAMMARY GLANDS*, or breasts, of the female. In fact, mammals get their name from these mammary glands which produce milk after the birth of a child.

The Male Reproductive System

The *TESTES* and the *PENIS* are organs of the male reproductive system. The testes produce the male sex cells called sperm cells. The testes are contained in a sac called the *SCROTUM*, which lies outside of the main body wall. Sperm are sensitive to heat, and this

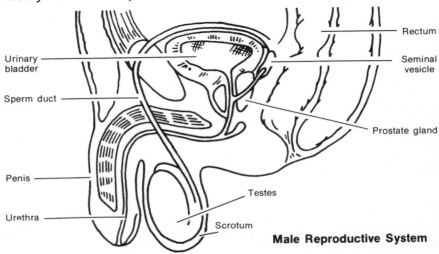

Male Reproductive System

location, outside the body proper, provides some relief from the higher temperature inside the body. Like humans, monkeys, horses, and cattle have testes which are outside the body. Other animals such as frogs, birds, and snakes have testes which are inside of the body cavity.

Sperm cells travel through a tube in the penis called the *URETHRA.* The *PROSTATE GLAND* also connects to the urethra. The prostate gland provides additional fluid called *SEMEN* which mixes with the sperm cells and carries them through the urethra. The semen then flows through the urethra, which travels through the penis to the outside of the body. Urine also flows through the urethra of the penis to the outside. However, urine and sperm do not flow through the urethra at the same time.

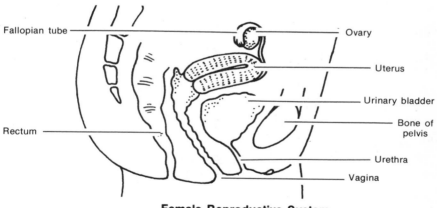

Female Reproductive System

The Female Reproductive System

Eggs are produced in *ovaries*, as in all other animals. The two ovaries are located inside the abdomen, below the stomach. One egg cell is produced at a time, usually about every 28 days. The exact time varies with the individual, but *OVULATION*, or egg production, occurs about once each month.

From the ovary where it is formed, the egg cell travels through a tube into the uterus, or womb. If sperm are present, it is fertilized in the tube. If it is not fertilized, the egg dies and passes out of the body.

Menstruation

Each month, the uterus prepares to receive a zygote. Its walls thicken to form a soft, blood-rich cushion waiting to hold and nourish an embryo. If the egg cell is not fertilized, the lining of the uterus breaks down and is discharged from the uterus through the vagina. Most of the discharge is blood that was in the tissues of the lining. This process is called *MENSTRUATION* and is necessary so that a fresh lining can be prepared for the next egg cell a month later.

Human Fertilization

Like all mammals, humans use internal fertilization. The penis, the male reproductive organ, is used as the means of getting sperm to the egg in the body of the female. In many species, the penis is concealed in a small cavity, but in humans it is external.

In preparation for fertilization, blood flows into special vessels in the penis, extending its length and making it rigid. Only when erect can it be inserted into the female's vagina, an opening which leads to the uterus. Semen, which contains sperm, together with fluid from the prostate gland is released into the vagina. Sperm cells must then swim to the uterus under their own power. If an egg cell is present, one of the sperm cells breaks through the egg membrane to fertilize it. The female is now pregnant.

Pregnancy

If the egg cell is fertilized, the wall of the uterus catches it and gives it a home. The embryo forms a placenta, which meshes with the *uterine wall* like a hand with thousands of fingers inside a glove made to fit it perfectly. The mother's blood flows through the uterine lining, and the baby's blood flows through the placenta.

The two blood supplies do not mix, but they come so close together that food and oxygen pass from the mother's blood to the baby's. Waste products go the other way, and the mother gets rid of them. Her blood does not flow through the baby directly, though.

The placenta is connected to the embryo by a cord called the *UMBILICAL CORD*, which contains blood vessels and carries blood between them. The embryo is well developed by the age of five weeks. The umbilical cord is tied off at birth and becomes the "belly button."

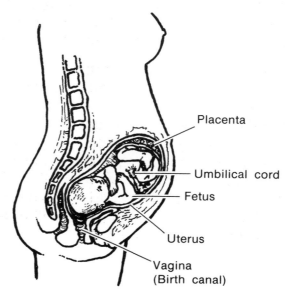

Pregnancy at Nine Months

The Embryo Becomes a Baby

Inside the uterus, the embryo develops rapidly, using the food supply provided by the body of the mother. The embryo first forms a hollow ball of cells. Quickly, the cells become different in appearance and function. The body takes shape, and organs appear. DIFFERENTIATION and DEVELOPMENT continue to change cells to tissues and tissues to organs.

After about three weeks, the heart begins to beat. Blood vessels form rapidly, and the body clearly shows a head, buds for arms and legs — and a tail. The embryo is still smaller than a fingernail.

The human embryo at three weeks looks just like the embryo of a fish, chick, or horse. (All animals with backbones look very similar in their early stages of development.) A human embryo soon moves on from this primitive stage, absorbing the tail and gills — yes, gills, too! — and using the cells for other body parts.

At about four weeks, tiny hands begin to show fingers, and eyes start to show up as dark spots. By the time pregnancy is half over, at 4½ months, the FETUS shows thumb sucking behaviors inside the womb. (The embryo is called a fetus in the later stages of development.) Sucking is an innate behavior which the newborn baby uses to suck milk from the mammary glands.

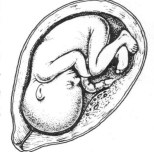

A fetus at 6 months

Birth of a Baby

When the fetus reaches full size, most often after nine months of development, the uterus begins to contract. Over and over, the uterus squeezes to push out the baby. Slowly at first, every half hour or so, then gradually faster until contractions come about every five minutes, the process of delivery goes on.

After much work, the mother has a new baby in her arms. The umbilical cord is still attached at the navel, and the placenta is pushed out soon after the baby is born. The cord is clamped and cut, and the baby can live outside of the mother's body.

Parental Care

The parents care for and feed their new offspring for many years. Compared to a baby horse, which stands in a few hours, runs the next day, and stops nursing after only a few months, the human baby is still helpless. There is no chance that human offspring could fend for themselves in less than several years.

Much time and effort must be devoted to caring for babies and young children. The basis of the human family is the caring and nurturing of children. Other animals have hundreds of thousands of offspring at a time and then leave them without care. Humans are different. Human beings usually have one child at a time and often several years apart. The care that they give their children is one reason for the high survival rate of humans.

Adolescence

Humans care for their offspring through the special period of the teens called *ADOLESCENCE* and often to adulthood. Adolescence is a period of increased development in humans. No other species has such an extensive and lengthy period of growing up. Growing up brings with it some special and rapid physical changes. These changes take place during *PUBERTY*, which occurs at the beginning of adolescence.

Changes in Males and Females

In the cases of males, there is a change in the voice from a high to a low pitch. Hair begins to grow on the face, under the arms, and in the *PUBIC* region where the sex organs are located. The sexual organs, themselves, become more fully developed, and the production of sperm is highly increased.

Females also undergo changes during puberty. Pubic and underarm hair start to grow. The mammary glands enlarge, and the menstrual cycle begins. Adolescents are fully capable of reproduction.

But, not only is adolescence a period of physical growth, it is also a time when there is increased emphasis upon individual behavior and the assumption of responsibility. In humans, the period of adolescence is the developmental stage in which the child becomes more and more like adults — and very often more like the parents.

CHECK YOUR UNDERSTANDING

1. What is the purpose of mammary glands?

2. What gland provides the fluid for the transportation of sperm?

3. Describe the path followed by sperm from the testes to the site of fertilization.

4. What is menstruation?

5. What does a three-week-old human embryo look like?

6. What are some changes that occur in males during puberty?

SUMMARY OF CHAPTER 9

A species becomes extinct if it does not reproduce. Reproduction insures that traits of the parents are passed on to the next generation. The traits of an individual lie in the chromosomes of the cells. Chromosomes contain DNA which gives the cell and the individual its hereditary qualities and characteristics.

Cell Duplication

Cells generally reproduce by mitosis. In this process the cell first duplicates its DNA. Then it divides in half. The daughter cells are identical to the parent cell.

In sexual reproduction, two parents are involved. In a process called meiosis half of the genetic material comes from each of the parents. This material goes to the egg cell and to the sperm cell. An advantage of sexual reproduction is that all of the genetic material does not come from just one parent. It creates great diversity among offspring. The offspring cannot be exactly like either parent.

Reproduction, Growth, and Development

Although there are many different animals in the world, all species follow similar patterns of growth and development. When a sperm cell goes through the membrane of an egg cell, a zygote is formed. A zygote is a fertilized egg. It begins as one cell, but divides and differentiates, until it is an embryo with body parts and organs.

In some species, the female lays eggs, and the eggs are fertilized outside her body. In mammals, the egg is fertilized internally and develops inside the body of the mother. Since the mammal nurtures and protects the developing embryo, more individual mammals survive and are therefore more successful than other species.

Human Reproduction

Like other mammals, humans fertilize their egg cells internally. Developing embryos are protected in the uterus and are nourished from the mother's body. After birth, babies are fed milk and are cared for by their parents for many years. In humans, there is a period of development called adolescence, in which children gradually become adults.

WORDS TO USE

1. extinct
2. chromosome
3. spontaneous generation
4. adapt
5. diversity
6. gene
7. DNA
8. resistant
9. sexual reproduction
10. egg cell
11. gonad
12. fertilize
13. external fertilization
14. embryo
15. nucleus
16. asexual reproduction
17. sperm cell
18. gamete
19. zygote
20. mitosis
21. ovary
22. testes
23. meiosis
24. membrane
25. reproduce
26. internal fertilization
27. gestation
28. uterus
29. cell differentiation
30. placenta
31. marsupial
32. scrotum
33. penis
34. semen
35. development
36. urethra
37. ovulation
38. vagina
39. adolescence
40. mammary glands
41. prostate gland
42. womb
43. umbilical cord
44. puberty
45. menstruation
46. fetus
47. pubic

• REVIEW QUESTIONS FOR CHAPTER 9 •

1. Why must living organisms reproduce?

2. Why do brothers look different from each other?

3. What is DNA and why is it important?

4. How does adaptation give a species a greater chance for survival?

5. How is DNA copied?

6. Why do insecticides only work well for a few years?

7. What are chromosomes?

8. What are genes?

9. Describe the process of mitosis.

10. What are gonads?

11. What is the difference between mitosis and meiosis?

12. What is an example of asexual reproduction in animals?

13. What is the advantage of sexual reproduction over asexual reproduction?

14. Where does meiosis occur?

15. What is the difference in chromosome number in gametes and other body cells?

16. What is the difference between a zygote and an embryo?

17. What is cell differentiation?

18. Describe how external fertilization occurs in fish.

• REVIEW QUESTIONS FOR CHAPTER 9 •

19. What is the difference between the way frog babies and human babies develop?

20. Why must fish lay hundreds of eggs?

21. What is gestation time?

22. What is a placenta?

23. How is fish reproduction different from human reproduction?

24. Why can only one sperm fertilize an egg?

25. Name two ways that humans are like other mammals.

26. Why do humans produce relatively few babies?

27. What evidence is there in the development of an animal embryo that many animals have a common background?

28. How does a baby get out of the uterus?

29. What is the belly button?

30. Why are human testes outside the body wall?

31. What happens to a human egg that is not fertilized?

32. What is the difference between internal and external fertilization?

33. What is ovulation?

34. What are some changes that occur in females during puberty?

Chapter 10

Understanding The Human Fight To Stay Healthy

Chapter Goal:

To discover how disease and diet each have a part in the human fight to stay healthy.

Key Ideas:

- The human fight to stay healthy is the story of progress in the control, prevention, and cure of diseases that affect the body.

- Good nutrition and a balanced diet are major factors in the human fight to stay healthy.

> **Key Idea #1:**
>
> - The human fight to stay healthy is a story of progress in the control, prevention, and cure of diseases that affect the body.

THE HUMAN FIGHT AGAINST DISEASE

Infectious Diseases

At one time, *INFECTIOUS* diseases were responsible for wiping out whole populations. In the 1300's, two-thirds of the people of Europe were stricken by the bubonic plague, or Black Death, as it was called. In the year 1665, more than 30,000 people in London died of this disease. In the 18th century, one out of every three English children was infected with smallpox, and those who survived were scarred with pockmarks. In the 19th century, a plague of another dread disease, cholera, wiped out thousands of people in the United States. Even in the twentieth century, polio, a disease of the nervous system, was a killer and crippler of children. You probably will not get these diseases. In the United States and other industrialized countries, these diseases have been practically eliminated.

Germs

At one time people thought that the wind or evil spirits were responsible for bringing infectious diseases that killed whole populations. Now it is known that *GERMS* or *PATHOGENS* cause disease. Pathogens are disease-causing materials. They are usually so small that they can only be seen with a microscope. For this reason they are called *MICROBES*. Microbes can be protozoans, bacteria, or viruses. These *MICRO-ORGANISMS* can be carried from place to place by mosquitoes, flies, fleas, rats, and pigs, and then transmitted to humans. Microbes are found in water, waste products, unwashed silverware, and in the coughs and sneezes of people who are sick. Microbes are everywhere. If microbes are everywhere, why aren't you sick all of the time?

The Body's Defense Systems

You stay healthy because your body has a series of defense systems that prevent germs from entering. The first lines of defense in the body are the skin, the lining of the nose and throat, the hairs of the nostril, the tonsils, and acids in the stomach. If the microbes get past these first lines of defense, they are surrounded by white blood cells. The white blood cells capture the germs before they can do any harm. Even with those two defense systems, some microbes do get into the cells. What happens then? That's when the body's *IMMUNE* system takes over.

Antibodies

Have you ever heard someone say, "Oh, I can't catch measles. I'm immune to them. My brother had them last year, and I slept in the same room, but I never caught them." That person probably has a *NATURAL IMMUNITY* to measles. You might have heard another person say, "I'll never get mumps, I just had a mumps shot." The second person has an *ACQUIRED IMMUNITY* to mumps. He or she has been *VACCINATED* or *INOCULATED* against mumps.

Your immune system system is your most important defense against disease. When a foreign substance enters your body, your blood forms *ANTIBODIES* which help combat the disease. These antibodies help you recover and prevent future attacks. When you get a "shot," you are inoculated with small amounts of a disease. This inoculation makes your body produce antibodies to fight the disease. When your body has produced antibodies against the disease, you will not be able to catch it. You have been *IMMUNIZED* against this disease. You can say that you have an *ACQUIRED IMMUNITY* because of the shot.

Vaccination

At one time, smallpox was a dreaded disease. In the early 1800's, people were inoculated against smallpox by being injected with a small amount of the disease. Unfortunately, some people died as a result of this inoculation. An English doctor, Edward Jenner, noticed that milkmaids who caught cowpox, a mild illness, never developed smallpox. He got the idea that if he introduced a small amount of cowpox into a person's body, that person would be immune to smallpox.

Jenner put cowpox material taken from a cowpox sore into the arm of a young boy. The boy developed cowpox and recovered. Next, Jenner inoculated the boy with smallpox. The boy did not get smallpox at all. The cowpox material had caused the boy's body to make antibodies. The boy was immune to smallpox. Jenner called his method *vaccination*. Some of you might have a small scar on your body, sometimes on the upper arm. This is a vaccination scar. Today, because of Jenner's work, smallpox has been eliminated from most parts of the world.

Vaccinations have helped people to become immune to smallpox, measles, mumps, polio, and other infectious diseases.

Sanitation

If you are thirsty, you can get a drink of water. You don't even think about it. You just turn on the faucet and get a cold drink. But at one time this was not the case. People drank water that was piped in from rivers in which filth and sewage were present. Or they got their water from town pumps where the water was CONTAMINATED by cesspools. Waste products in the cesspools and rivers filled the water with bacteria and viruses. There were EPIDEMICS of typhoid fever and cholera. Thousands of people grew ill and died. Better sanitation methods and practices are the leading cause of improved health and chances to stay alive.

No one at that time knew or thought much about microbes. They did not realize that germs in water or food could infect masses of people. Even medical doctors did not know that germs could travel from one patient and infect the next patient. Louis Pasteur, a French scientist, was one of the first scientists to study microbes.

Today, people know that microorganisms can cause disease. They are careful to keep things clean and sanitary. For that reason they do not drink water or eat food that has been contaminated by bacteria. The water you drink is filtered and purified before it ever reaches the faucet. It is tested by government agencies to make sure it is fit to drink. Food in the supermarkets and medicines in the drug stores have to be safe before they can be sold. Milk has to be pasteurized, or heated in a special way, to make sure the bacteria are destroyed.

CHECK YOUR UNDERSTANDING

1. Name some microbes that cause infectious diseases. Tell how these diseases are transmitted.
2. What are the first two lines of defense of the body against disease?
3. What are antibodies? How do they fight disease?
4. What is being done to make sure that our water is fit to drink?

Key Idea #2:

- Good nutrition and a balanced diet are major factors in the human fight to stay healthy.

NUTRITION AND DIET

People often say, "Good nutrition is important if you want to stay healthy." Nutrition is all about how the body uses food. Foods are *NUTRIENTS* because they *NOURISH* the body or feed it. Nutrients are assimilated and used by the cells.

Your cells function properly when they have the substances that they need. These are: carbohydrates, fats and oils, proteins, water, *VITAMINS*, and *MINERALS*. Each of these substances is used in a special way by the body. Carbohydrates are made of sugars and starches and are the energy producers. That's why you have so much energy after you eat a candy bar, which is full of sugar. Fats and oils give you energy. Proteins are tissue builders. This is the material that the cells mostly use to repair themselves and to reproduce.

Water makes up about 70% of your body tissue. It is the material in which all the body substances are dissolved. You lose water when you sweat. You also lose water when you get rid of body wastes during urination. Your body needs a great deal of water. You need to ingest enough liquids so that you won't *DEHYDRATE,* or dry out.

Vitamins

Vitamins are substances that are found in tiny amounts in animal and plant foods. The vitamin chart on the next page shows how essential vitamins are.

ESSENTIAL VITAMINS

Vitamin and Source	What It Does	Deficiency Disease
Vitamin A Milk, egg yolk, beef liver	Helps skin, hair, eyes, and lining of nose and throat; prevents night blindness	Dry, rough skin, lowered resistance to respiratory infections, poor vision in twilight.
Niacin Liver, meat, whole wheat, milk	Protects skin and nerves, aids digestion.	Mental depression, digestive disturbances
Vitamin B (Thiamine) Milk, pork, egg yolk, cereal, vegetables, fruit	Protects health of nervous system, aids appetite and digestion.	Tiredness, loss of appetite.
Vitamin B$_2$ (Riboflavin) Milk, pork, liver, eggs, vegetables, fruit	Increases body resistance to infection and prevents harmful changes to eyes.	Eye fatigue, lower vitality.
Vitamin C (Ascorbic Acid) Tomatoes, most citrus fruits	Helps form bones and teeth; prevents scurvy.	Scurvy, sore joints, tender gums, poorly formed bones and teeth.
Vitamin D Fish liver oils, milk, sunlight	Prevents rickets, uses calcium and phosphorus to build bones and teeth.	Rickets, poorly formed bones and teeth
Vitamin E Vegetable oils, cereals, eggs.	Probable role in reproduction.	Not known.
Vitamin K Green leafy vegetables, soy beans, bran.	Aids in clotting of blood.	Bleeding.

Minerals

Minerals are essential to good health. They are found in very small amounts in the foods that you eat. *Calcium* and *phosphorus* build strong bones and teeth. *Sodium* and *potassium* help the cells to function. Most people get enough minerals in the foods that they eat. But good food must be eaten each day, since minerals are not stored in the body.

Some people, especially older ones, have too much sodium. Sodium is table salt, and, as people grow older, they are often told to cut down on salt. They might need a "salt-free" diet, especially if they have high blood pressure. Although exact causes of high blood pressure are unknown, many doctors regard salt as a major factor.

The Four Basic Food Groups

Doctors suggest that you eat some foods from each of the four basic food groups. Your body will get the necessary nutrients if you select food from each of the four food groups on a daily basis.

FOUR BASIC FOOD GROUPS

Food Group	Foods in Group	Daily Requirement
Milk Group	Milk, butter, cheese, ice cream	4 glasses or their equivalent
Meat Group	Meats, poultry, eggs, peas, dry beans, nuts	2 or more servings
Vegetable-Fruit Group	Dark green and yellow vegetables, citrus fruits, tomatoes	4 or more servings
Bread-Cereal Group	Bread, cereals, crackers, spaghetti	4 or more servings

SUMMARY OF CHAPTER 10

Infectious diseases are caused by pathogens. Most pathogens are so small that they can only be seen under a microscope. The pathogens are called micro-organisms or microbes. Microbes can be bacteria, viruses, or other protozoans. At one time, microbes were responsible for widespread death and disease.

Doctors began to realize that microbes were the cause of disease. For this reason they advised people to drink clean water and to live under sanitary condi-tions. In time, it was discovered that people would develop an immunity to a disease if they were given a small amount of the disease itself. Inoculation caused the production of antibodies to fight the disease. Because of mass inoculation, most people are now immune to infectious diseases.

Nutrition is the process of ingesting, digesting, absorbing, and assimilating food. Nutrients in food are necessary for cell repair and growth. In order to function, the body needs carbohydrates, fats, proteins, water, vitamins, and minerals. A balanced diet supplies these substances. Eating food from the four basic food groups (meat, milk, grain, and fruit and vegetables) gives the body the nourishment it needs.

WORDS TO USE

1. immunization
2. epidemic
3. disease
4. inoculate
5. pasteurized
6. sanitation
7. vaccination
8. antibody
9. contaminated
10. infectious
11. microbe
12. pathogen
13. nourish
14. dehydrate
15. vitamin
16. mineral
17. nutrition
18. nutrient

• REVIEW QUESTIONS FOR CHAPTER 10 •

1. What are pathogens?

2. What is a microorganism? How can you see one?

3. How do blood cells prevent you from getting sick?

4. What is the difference between a natural immunity and an acquired immunity?

5. How did Edward Jenner fight disease?

6. Who was Louis Pasteur?

7. How do inoculation and sanitation prevent infectious diseases?

8. What is nutrition?

9. List four of the six substances that your cells need in order to function properly. Tell why each substance is necessary.

10. Doctors do not agree with each other about taking vitamins. Why don't they agree?

11. How can salt affect blood pressure?

12. Why is it important for you to eat proteins?

13. Name the four basic food groups. List two foods from each of the four groups.

Chapter 11

Following Genetics From Generation to Generation

Chapter Goal:

To examine and understand how traits and features of one generation are passed to the offspring in the next generation of organisms.

Key Ideas:

- The transmission of traits from one generation to the next occurs in a regular pattern.

- The basis of heredity is what chromosomes do during sex cell formation.

- Human genetics is the study of heredity as it applies to humans.

- Applied genetics makes it possible to change the traits of organisms.

INTRODUCTION

Heredity

The way you resemble your parents, from the freckles on your nose to the length of your legs to your skill in arithmetic is due to your *heredity*. The laws of heredity apply to all organisms, whether they are rose bushes or spiders or frogs. Special cells from parents contain the chemicals that determine what a new individual will be like.

The human species, by means of sexual reproduction, produces unique, one-of-a-kind, persons. Think of the great variety of human beings around you. You can see that each person is different from the next person. Fingerprints are one of a kind. Newborn babies have their footprints made at the hospital to identify them.

Environment

Heredity does not stand alone. There is always an *ENVIRONMENT* which affects the original material. A tomato seed may have inherited the ability to produce the greatest tomatoes in the world. But, without an environment of sunshine, proper soil, water, and favorable temperature, it will not become as good a tomato plant as it could have been.

Genetics

GENETICS is the study of heredity. It is all about the basic *laws of heredity* and how these laws work. By studying genetics, you can find out about human heredity and how you came to be you. Genetics also includes the science of improving plant and animal species.

> **Key Idea #1:**
>
> * The transmission of traits from one generation to the next occurs in a regular pattern.

WHAT IS HEREDITY?

Family members have some of the same *TRAITS*. The passing of traits from parents to their children is called *HEREDITY*.

Have you ever been told that you look like someone in your family? Have you ever observed that people in a particular family look alike?

All organisms pass their traits to their children or *OFFSPRING*. Racehorses look like other racehorses. Plow horses look like other plow horses. Calico cats have calico cats. Grey cats have litters of grey kittens.

Gregor Mendel

Over one hundred years ago, a monk named Gregor *MENDEL* made some important discoveries about heredity from his experiments with pea plants. Mendel took seeds from tall pea plants and planted them. Most of the plants from the "tall" seeds were tall, but some of them were short. Mendel wanted to find out why the short pea plants were unlike their parents.

Pure Traits

Mendel began his investigation by growing plants that were *PURE* for the trait of tallness. Plants that were pure for this trait would only produce tall plants from their seeds. He did this by the method of *SELF-POLLINATION*. Flowering plants have both the male and female sex organs in their flowers.

Mendel transferred the pollen from the *ANTHER,* the male sex organ, to the *PISTIL*, the female sex organ. In this way, the ovary could be fertilized, and seeds would be produced. Mendel used these seeds to grow pure, tall pea plants.

Using the same method of self-pollination, Mendel grew plants that were pure for the trait of shortness. He called the "pure tall" and the "pure short" plants the P_1 *GENERATION*. P_1 stood for the first or parent generation of pure plants. (To understand what a generation is, think of your grandparents as one generation, your parents as another generation, and yourselves as still another generation.)

The Next Generation
Mendel wanted to find out what a pea plant would be like if it had one pure tall parent and one pure short parent. He used the method of *CROSS-POLLINATION* to provide two different parents. Mendel transferred the pollen of pure tall pea plants to the pistils of pure short pea plants. He also did the reverse. He took the pollen of pure short plants to fertilize the ovaries of pure tall plants.

Then Mendel took the seeds from the crossed plants and planted them. He called the new plants the F_1 generation. The "F" stood for filial, or son.

To his surprise, Mendel found that all the new plants in the F_1 generation were tall. Mendel crossed thousands of pure short and pure tall plants. He observed the seeds and the new plants carefully. But, each time, the results were the same. All the plants were tall. What had happened to shortness? Why had this trait disappeared?

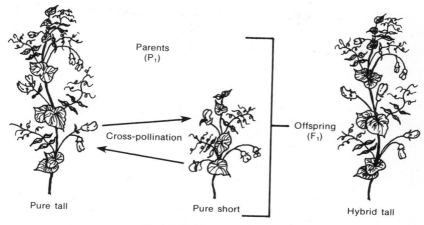

Mendel's Experiment

Punnett Squares

A special arrangement called a *PUNNETT SQUARE* can be used to show Mendel's findings. A capital T is the symbol for the trait of tallness. A small t is the symbol for the trait of shortness. TT stands for a pure tall pea plant, and tt stands for a pure short pea plant.

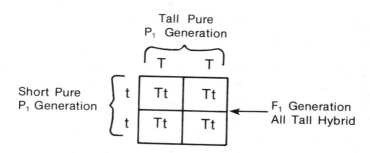

When Mendel crossed pure tall pea plants with pure short pea plants, he produced an F_1 generation of all tall plants. You can see by the symbols in the boxes that shortness did not disappear. It is still there. But it does not show up, because shortness is a factor of less influence.

Mendel's factors are called *GENES*. A factor that is often hidden is a *RECESSIVE* gene. A factor that *does* show up is called a *DOMINANT* gene. A recessive gene does not produce an effect when with a dominant gene. Plants that have more than one kind of gene for the same trait are called *HYBRID* plants. The plants in Mendel's F_1 generation were hybrids. They had a gene for tall and a gene for short.

Genotypes and Phenotypes

A hybrid has two different genes. One gene has a stronger influence, and the other gene has less influence. The pea plants in the F_1 generation were tall, because tallness was caused by the gene partner with the stronger influence. However, inside the pea plants there was a combination of tall and short genes in pairs. What is inside an organism, in its genes, is called the *GENOTYPE*. What the organism looks like — its appearance — is its *PHENOTYPE*. All organisms have both.

The F_1 Generation

Mendel worked in his monastery garden and produced another generation of pea plants by self-pollinating the F_1 generation. He called this third group of plants the F_2 generation. He went from P_1 to F_1 and now F_2, three generations of painstaking gardening and record keeping. This time, Mendel found that shortness had returned. In the F_2 generation, 3/4 of the plants were tall, and 1/4 were short. Mendel repeated his experiments and always got the same results. The percentage in the F_2 generation was always 3/4 tall and 1/4 short.

The Punnett square on the next page shows what the F_2 generation was like. Three fourths of the plants were tall, and one fourth were short.

What percent of the plants were pure? What percent were hybrid?

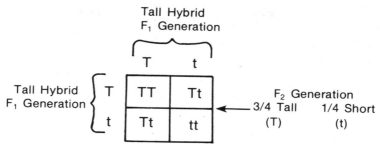

Tall Hybrid
F₁ Generation

Tall Hybrid
F₁ Generation

F₂ Generation
3/4 Tall   1/4 Short
(T)   (t)

Mendel's Laws

Mendel also checked other traits such as seed color, seed texture, pod appearance, pod color, and position of the flowers. Each trait followed the same kind of pattern. Round seeds were dominant over wrinkled seeds. Yellow seeds were dominant over green seeds. Green pods were dominant over yellow ones. Mendel concluded that there must be some *factor* in a plant that causes it to have certain traits such as tallness or shortness. He reasoned that since *two factors* are involved, they must come in pairs. Mendel also thought that one factor must be more powerful than the other factor, since the factor of stronger influence sometimes hides the factor of less influence. Mendel called the stronger factor "dominant" and the weaker factor "recessive."

Mendel also thought that the paired factors must separate during reproduction, so that the new egg gets half the factors from one parent and half from the other parent. Today, *these factors are called genes*, and the study of heredity is called genetics. In Mendel's experiments, genes were responsible for producing short or tall pea plants. Genes give you your blue eyes or your brown hair. Genes are responsible for carrying traits from one generation to the next.

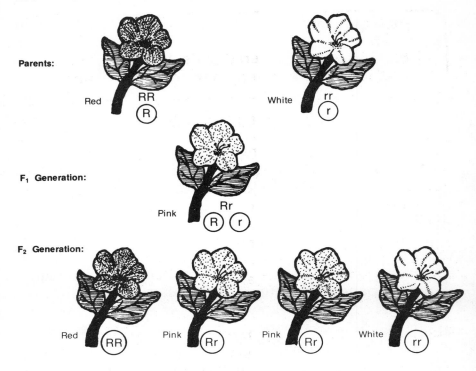

Parents:

Red   RR   White   rr
R   r

F₁ Generation:

Pink   Rr
R   r

F₂ Generation:

Red   RR   Pink   Rr   Pink   Rr   White   rr

How Color is Determined in Certain Flowers

CHECK YOUR UNDERSTANDING

1. What is heredity?
2. What is a pure trait?
3. What did Mendel do to produce a plant that had one tall parent and one short parent?
4. What were the dominant and recessive factors that Mendel found? Why did he call them dominant and recessive?
5. What is genetics?
6. What is a hybrid?
7. What is the genotype and the phenotype of a Tt plant?

Key Idea #2:

- The basis of heredity is what chromosomes do during sex cell formation.

WHAT CHROMOSOMES DO

CHROMOSOMES are tiny, rod-shaped bodies located in the nucleus of each cell. Chromosomes carry hereditary information. Genes are located on the chromosomes. There are gene pairs for every trait in an organism. Heredity is determined by the way the chromosomes divide during the formation of gametes or sex cells.

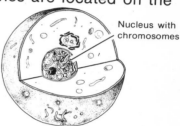

Nucleus with chromosomes

Body Cell

Mitosis and Cell Division

From your study of simple organisms, you know that amebas reproduce by splitting in half. The nucleus of the ameba has chromosomes that duplicate. During division of the nucleus, two sets of chromosomes are made. The process is called *MITOSIS*. The new amebas each get a set of chromosomes and are exactly like their parent. The ameba is able to duplicate itself because the chromosomes in its nucleus go through mitosis. The whole cell also divides.

In the process of mitosis, first the chromosomes in the nucleus duplicate themselves. Then the nucleus membrane dissolves, and half the chromosome materials go to form one new nucleus. The other half forms a second nucleus. The cell itself divides, and each cell then has a nucleus with a complete set of chromosomes and genes exactly like the original set. Since the chromosomes are the same, the new amebas are the same. In amebas, parent and offspring are look-alikes.

Not only do one-celled animals carry out the process of mitosis and cell division; higher organisms do, also. Body cells are constantly dividing as you grow. But, in higher organisms, *body cells* divide by mitosis and cell division only for growth.

Higher organisms reproduce sexually. Two parents are involved. Each parent contributes sets of chromosomes in a male or female *GAMETE*. Gametes are egg and sperm *sex cells*.

Reduction Division

In sexual reproduction, half the chromosomes come from one parent, and half come from the other parent. If sperm sex cells and egg sex cells just followed the pattern of regular cell division, they would have double sets of chromosomes in their nucleuses. Then when fertilization of the egg cell occurred, the beginning organism would have *two* sets or *twice as much genetic material as needed* by the species. (Remember, each species has a set number of chromosomes.) If carried to the next generation, it would even be more irregular. So *that doesn't happen at all*.

Doubling of chromosomes for sperm sex cells and egg sex cells is *prevented* by a special process of *reduction* during division. *REDUCTION DIVISION* occurs during *MEIOSIS* (my oh sis), or sex cell formation. Reduction division takes place *only* in the formation of eggs and sperms. The process involves several steps.

Each sex cell divides, so that it contains 1/2 of the species number of chromosomes. When a *ZYGOTE* is formed by the *union of the sex cells*, their half number (half the chromosomes from each parent) becomes a whole (1/2 + 1/2) number that is the species number.

Human beings have 46 chromosomes — twenty-three pairs. If the 46 chromosomes in the sperm cell got together with the 46 chromosomes in the egg cell, a beginning individual would have 92 chromosomes, twice as many chromosomes as the normal species number. Because of reduction division, the sperm cell contains 23 chromosomes and the egg cell contains 23 chromosomes — one-half the species number. When a new embryo is formed, it then has 46 chromosomes, 23 from each parent. Forty-six is the normal number of chromosomes for all humans.

Because of the mix of chromosomes from two parents, sexual reproduction produces a completely different organism. The combination of genes that is made from the genes of two parents is a major feature of sexual reproduction. Unlike the ameba, organisms resulting from sexual reproduction are *one of a kind*.

Examples of pairs of chromosomes of a human male

Mendel's Work Rediscovered

By the beginning of the 20th century, chromosomes could actually be seen under a microscope. Scientists had not paid much attention to Mendel's experiments before this. Now they realized what his work had proved. They concluded that what Mendel had called *factors* were the *genes* that passed traits from one generation to the next. At this time, biologists also reasoned that one chromosome came from the female parent and the other from the male parent, and that there were opposite forms of a gene, called *ALLELES*, such as tall or short for a pea plant. This was called the chromosome theory.

Sex Chromosomes

Biologists can observe the *SEX CHROMOSOMES*. Humans have 23 pairs of chromosomes for a total of 46. In 22 of the pairs, there are two chromosomes that look exactly alike. But there is one pair of chromosomes that is different in males and females. These are called the *sex chromosomes*. Sex chromosomes determine the sex of an individual. They give male or female characteristics to people, such as a beard and a lower voice to men.

In mammals and flies, the female sex chromosomes match. They are called the X chromosomes, and are represented by the XX symbol. In males, the chromosomes do not match. They are called the XY chromosomes.

The Punnett square at the right shows how the sex chromosomes determine the sex of an individual. The chances are 50/50 of producing either a female (XX) or male (XY).

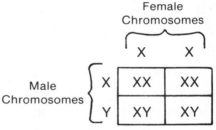

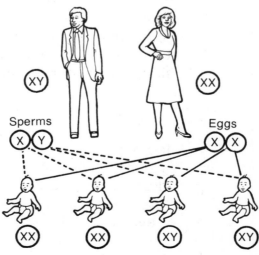

Experiments with Fruit Flies

DROSOPHILA, a fruit fly with red eyes, has been the subject of many genetics experiments. The *Drosophila* is a good subject for many reasons. Its cells have four pairs of chromosomes, and they are easy to see, since they are huge for such a small fly. *Drosophila* reproduces quickly, is inexpensive to feed, and can be raised in small jars or vials. In addition, it is easy to tell the male from the female. The two-name system holds true for these main players in the study of genetics. They are *Drosophila melanogaster*. No other flies admitted!

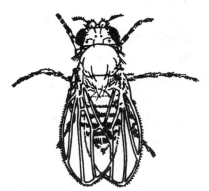

Drosophila melanogaster

Thomas Morgan, a scientist famous for his study of genetics, used *Drosophila* for his experiments. He noticed that one of the male flies had white rather than red eyes. He mated a white-eyed male parent with a red-eyed female parent in order to produce an F_1 generation. All the offspring in the F_1 generation had red eyes. Morgan concluded that red eyes were dominant (Rr) in fruit flies. Mendel had reached the same kind of conclusion about tallness in pea plants (Tt).

Sex-Linked Traits

Next, Morgan mated the flies from the F_1 generation to produce an F_2 generation. Again, the results were the same as Mendel's. Except for one astonishing factor! *All the white-eyed flies were males!* Morgan concluded that white eye color in *Drosophila* is a *SEX-LINKED TRAIT*.

Females have a pair of X chromosomes, and are represented as XX. The gene for white eyes is located on the X chromosome from the female. There is no allele for eye color on the male chromosome (XY). The trait for white eyes can only be passed by a female to a male. That is why it is called a sex-linked trait. An organism that has a trait in its genes and passes it on to another generation is called a *CARRIER*. Being a carrier, the female *Drosophila* never has white eyes herself. However, she can pass down this trait to her male offspring.

CHECK YOUR UNDERSTANDING

1. What are chromosomes?

2. What is mitosis?

3. What is reduction division?

4. What are alleles?

5. What is the difference between the sex chromosomes and other pairs of chromosomes? Write the symbols for the female and male sex chromosomes.

6. Why was the *Drosophila* used for many experiments with genes?

7. In what ways were the results of Thomas Morgan's experiments like the results of Gregor Mendel's experiments? In what way were the results different?

Key Idea #3:

- Human genetics is the study of heredity as it applies to humans.

HUMAN GENETICS

Studying human genetics presents more problems than studying heredity in a fruit fly. 1) Human beings have more chromosomes than a fruit fly. 2) Humans do not reproduce as quickly, and 3) Humans cannot be used in laboratory experiments. Scientists study human heredity by *observing* how heredity and environment affect identical twins. They look at whole families and see how genetic diseases are passed from one generation to the next.

The Gene Pool

To study human genetics, scientists look at samples from the *GENE POOL*. The gene pool is all the genes in a certain population, such as the population of the United States, or all the genes in a population of white-tailed deer. Scientists can observe specific traits in one generation. These traits are likely to show up in the same proportion in the next generation. A large number of people immigrated to the United States from a great variety of places, so the gene pool of this country is large and varied. Other countries have kept a more uniform population. Their gene pools are less varied than that in the United States.

Studies of Twins

Scientists observe twins to study human genetics. There are two types of twins, *IDENTICAL TWINS* and *FRATERNAL TWINS*. Identical twins have duplicate genes, because they come from a single embryo. First, the sperm cell fertilizes the egg cell. Then the zygote divides into two separate cells. Two identical organisms are produced.

Fraternal twins are not identical. They are just like ordinary brothers and sisters. *Two different eggs* are fertilized by two different sperm cells at the same time. The result is two different offspring with completely different sets of genes.

Heredity and Environment

You are born with certain genes. That is your heredity. Your genes give you your skin color, your intelligence, your basic body structure, and other basic attributes. But, you live in an environment that contains all the things around you. Your family, your school, your friends, your total surroundings, all make up your environment. *GENETICISTS* often study identical twins who have been separated since birth. Both twins have the same genes, but they have grown up in different environments. Geneticists make comparisons and observe the effect of heredity and the influence of environment on a person's development.

The Influence of Environment

Poor nutrition, drugs and alcohol, and smoking can affect an individual. The same factors in the internal environment during pregnancy can affect the development of an embryo. Studies show that babies born to mothers who smoke cigarettes have *lower birth weights* than babies born to non-smoking mothers. Children of *heroin addicts* are addicts, themselves, when they are born. They have to be withdrawn from the drug after delivery. X-rays can also damage an embryo and cause changes or *MUTATIONS* in the genes.

DNA

Mutations can cause abnormalities in humans and other organisms. The white-eyed fruit fly was a mutation. But how do these genetic changes occur?

DNA is the substance in cells that contains the genetic material that is passed down from one generation to the next. All the information necessary to carry on the life activities is in the DNA molecules. The information is like a code that tells the cells what to do. A major feature of DNA is that it can *REPLICATE*, or duplicate, itself. Each DNA molecule has the ability to split or come apart. It can be compared to a "twisted ladder" that comes unzipped at the rungs (chemicals). New pieces of chemicals join each side of the ladder and an exact copy of the old ladder is made. Now there are two new DNA strands.

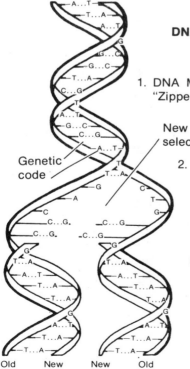

DNA Molecule

1. DNA Molecule, Showing "Zipper" Arrangment

New chemical materials are selected here.

2. When the zipper separates, each side takes on a new side. The new chemicals attach to the old sides and form new strands.

3. Each of these double strands contains one old strand and one newly replicated strand.

4. The rungs are the genetic code.

Genetic code

Old   New   New   Old

Because of this special ability to make copies, DNA is one of the most important chemicals in living things. Each individual has a separate DNA code. That is why you are different from every other individual.

Mutations

Sometimes there is a change in the DNA code. These are usually errors in the copying process. The change, called a *mutation*, can occur from no known cause, or it can be the result of an environmental factor such as exposure to radiation.

Some mutations are harmful, and others are beneficial. Many of these changes are small and do not have any great effect at all. Most mutations cannot even be seen, since the abnormal genes are recessive. However, if two parents with the same recessive genes marry, their children can inherit a *GENETIC DISEASE*.

Genetic Diseases

Diabetes. *DIABETES* is an inherited disorder caused by a recessive gene. In diabetes, the body is unable to manufacture *INSULIN*. Insulin regulates the amount of sugar in the bloodstream. People who have diabetes have an abnormal amount of sugar in their bodies. Diabetics can take artificial insulin, either by mouth or by injection, to lower the amount of sugar in their bodies.

Sickle-Cell Anemia. *SICKLE-CELL ANEMIA* is a genetic disorder which affects the red blood cells. The disease results from a pair of recessive genes that are inherited from parents who are carriers. The carriers have the trait for sickle-cell anemia in their genes, but do not have the disease, itself. The sickle-cell trait is carried by eight to ten percent of the black population in the United States. People with this disease have severe anemia which causes weakness and irregular heart action. At present, there is no cure for sickle-cell anemia.

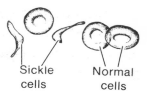

Sickle cells   Normal cells

Abnormalities Through Inbreeding

With the billions of people who live in the world, there are a great variety of genes. A person who is a carrier of a genetic disease is not likely to marry someone with the same genetic makeup. Their children will probably be normal. However, people within a small geographic area marry each other. Their offspring, again, marry within the area. So, there is a greater likelihood, after several generations, of their having children with genetic diseases. The same thing is true when people marry very close relatives. Marrying people within a small gene pool is called *INBREEDING*.

The Blue People. The blue people of Appalachia are examples of inbreeding. Blue people get their name because their genetic abnormality gives their skin a blue color. Aside from their looks, they are quite normal. The blue people originally lived in an isolated mountain village far from other towns. They married within their own group, simply because there was no one else to marry. As a result, their recessive genes frequently came to the surface.

Hemophilia. *HEMOPHILIA* is another example of a genetic disease caused by inbreeding. In hemophilia, the gene that controls blood clotting is defective. Hemophiliacs can bleed indefinitely, and need transfusions of normal blood so that their blood will clot.

Hemophilia is sometimes called the "Royal Disease," because many *kings and princes* in Europe suffered from it. People from royal families generally marry other people from royal families, including cousins. Since their gene pool is very small, recessive traits such as hemophilia show up in the next generation.

Sex-Linked and Sex-Influenced Traits

Did you notice that the last paragraph about hemophilia said, "many of the *kings and princes* in Europe suffered from it"? The sentence did not mention queens or princesses. Hemophilia, like white eyes in a fruit fly, is a sex-linked trait. It is carried by the female who passes it to the male offspring. *COLOR BLINDNESS* is another sex-linked trait. People with this condition cannot see certain colors. The most common type of color blindness is the inability to tell the difference between red and green. Mothers are *carriers* of color blindness, and pass the trait to their sons.

CHECK YOUR UNDERSTANDING

1. What is a gene pool?

2. What is the difference between heredity and environment?

3. What are some factors that can affect the development of an embryo?

4. What is DNA? What does it do?

5. What is a mutation?

6. What is a genetic disease? How is it passed from one generation to the next?

7. How does inbreeding increase the likelihood of inheriting a genetic disease?

> **Key Idea #4:**
>
> - Applied genetics makes it possible to change the traits of organisms.

APPLIED GENETICS

Plant and animal breeders use selective breeding to produce better organisms. Geneticists transfer desirable genes from one bacteria to another. Scientists are applying their knowledge of genetics to improve upon heredity.

The Survival of Useful Mutations

Are all mutations bad? Mutation may sound like a bad word, but it has caused many beneficial changes. Mutations in a population can increase the survival of a species of organisms. For example, a mutation that produces an animal with white fur would be helpful in a snowy region. Animals with white fur would probably live longer than those with dark fur, because they would blend in with the background. A plant population with a mutation that resisted certain weed killers would survive better if it were growing in an area where weed killer was used. Eventually, there would be more white-coated animals in northern regions and more weed-killer resistant plants. Over long periods of time, mutations cause changes in populations, and new species are formed.

Selective Breeding

Farmers and biologists use their knowledge of mutations to produce new varieties of plants and animals. Suppose a desirable mutation arises, such as sheep with short legs. Short legs may not seem to be a desirable trait in sheep. However, short legs prevent sheep from jumping over fences. This mutation actually did arise, and short-legged sheep were bred until a new generation of short-legged sheep was produced. The method of selecting useful mutations and breeding organisms so that the mutation shows up again is called *SELECTIVE BREEDING.*

Farmers use selective breeding to produce cows that give great amounts of milk and chickens that lay large numbers and large sizes of eggs. Selective breeding has also brought on the pink grapefruit and the navel orange.

Breeding Racehorses

Selective breeding is used to produce great racehorses. Good genes are what the horse breeders refer to when they talk about bloodlines in horses. Secretariat was a Triple Crown winner in racing in 1973. That means he won the Kentucky Derby, the Preakness, and the Belmont Stakes, the three most important races in the United States. In order to breed a winner, Secretariat's owner had selected parents with the right kind of traits that would produce a winner. Secretariat's father was known for fathering horses that won short races. From his mother's side, Secretariat inherited strength and the ability to go long distances. The result was Secretariat, an oversize horse with a stride of about 25 feet and with excellent speed. Breeders also talk about a horse having a good temperament. Behavior, too, is genetically controlled.

Racehorses are bred selectively.

New Techniques in Genetics

Scientists can examine chromosomes under the microscope and photograph them. They arrange them in pairs to see if there are any abnormalities. Sometimes they find a defective gene or an extra chromosome. As recently as the 1970's, scientists began to study genes by introducing new genes into an organism. This is called *GENETIC ENGINEERING*.

Biologists were first able to transfer genes from one species of bacteria to another species of bacteria. Now they are able to transfer genes between entirely different organisms. This discovery has enabled them to produce substances that are important in medicine, such as insulin, which is necessary for the treatment of diabetes.

So far, scientists have not been able to replace defective genes in humans with normal ones. However, research in this area is still ongoing. Guidelines for controlling genetic engineering experiments have been set by the scientists themselves.

CHECK YOUR UNDERSTANDING

1. What is a desirable mutation? Give an example of a mutation that is helpful.

2. What is selective breeding?

3. How did Secretariat's genes enable him to be a great racehorse?

4. What is genetic engineering?

SUMMARY OF CHAPTER 11

Heredity is the transmission of traits from one generation to the next. Heredity occurs in an orderly pattern. Many years ago, Gredor Mendel, while working with pea plants, discovered that traits in organisms are due to paired factors, half of which come from each parent. Mendel also found that some factors are dominant and others are recessive.

Mendel's factors are now called genes, and the study of heredity is called genetics. Genes are located on chromosomes which are found in the nucleus of each cell. Chromosomes are able to duplicate themselves, so that each new cell is an exact copy of the original. During the formation of sex cells, the sex cells divide in half. This process is called reduction division. Half of the chromosomes from each parent form the new organism.

Mutations are changes in the genetic material. Harmful mutations cause genetic diseases such as diabetes and hemophilia. Individuals can inherit a genetic disease if they receive the same recessive genes from both parents. Inbreeding causes an unusually large number of abnormalities. With inbreeding, the advantage that a large gene pool provides is lost.

Some mutations are helpful. In selective breeding, the results of mutations are used to produce better varieties of plants and animals. Today, scientists can transfer genes from one species to another. This process is called genetic engineering.

WORDS TO USE

1. environment
2. genetics
3. heredity
4. traits
5. offspring
6. Gregor Mendel
7. pure
8. self-pollination
9. anther
10. pistil
11. generation
12. cross-pollination
13. Punnett square
14. genes
15. recessive
16. dominant
17. hybrid
18. genotype
19. phenotype
20. chromosomes
21. mitosis
22. gamete
23. reduction division
24. meiosis
25. allele
26. sex chromosome
27. *Drosophila*
28. sex-linked trait
29. carrier
30. gene pool
31. identical twin
32. fraternal twin
33. geneticist
34. mutation
35. DNA
36. replicate
37. genetic disease
38. diabetes
39. insulin
40. sickle-cell anemia
41. inbreeding
42. blue people
43. hemophilia
44. color blindness
45. selective breeding
46. Secretariat
47. genetic engineering

• REVIEW QUESTIONS FOR CHAPTER 11 •

1. What are the factors that cause you to resemble your parents?

2. What kind of plants were produced when Mendel crossed tall pure pea plants with short pure pea plants? Why did Mendel get this result?

3. What were two of Mendel's four conclusions when he finished his experiments with the pea plants?

4. What is a Punnett square?

5. Write the symbols for:
    a) a tall, pure plant.
    b) a short, pure plant.
    c) a tall, hybrid plant.

6. What happens in the process of mitosis?

7. What happens in the process of meiosis?

8. What traits do sex chromosomes determine?

9. What was Morgan's conclusion when he found that only the male *Drosophila* had white eyes?

10. What are the problems in studying human heredity?

11. What is the difference between identical twins and fraternal twins?

12. Why do scientists use identical twins to study human heredity?

13. What are some environmental factors that can harm the embryo?

14. What is diabetes?

15. What is a carrier?

16. Why does a large gene pool help prevent genetic diseases?

Chapter 12

Tracing Cycles In The Environment: Ecology

Chapter Goals:

To find out how organisms interact with the living and non-living parts of their environment.

To discover how natural substances are used and recycled, and what happens to energy in the ecosystem.

Key Ideas:

- Ecosystems are sets of communities in which organisms interact with each other and with the non-living parts. Ecology is the study of ecosystems.

- Food chains and food webs are about the food and energy relationships in an ecosystem.

- All substances are constantly reused in a never-ending series of cycles as organisms carry out basic life activities.

- Energy is essential to all life. Energy constantly enters and leaves the earth and cannot be recycled.

INTRODUCTION

Living things depend on each other and upon the natural materials in the world around them. Air, water, and soil, along with energy from the sun, provide the basic needs of land plants and protists. Green plants and protists, in turn, provide for the basic needs of animals. When you study all of the living and non-living parts of ecosystems and how they relate to each other, you are studying *ECOLOGY.*

Studying ecology means learning about the environment and its interactions. It means studying the cycles in which natural substances are used and reused. It means understanding that food webs and food chains make living things work. Ecology also emphasizes the idea that energy from the sun is essential to all life. The main feature of ecosystems is the continued and high level of interaction between organisms and their living and non-living environments.

Key Idea #1:

- Ecosystems are sets of communities in which organisms interact with each other and with the non-living parts. Ecology is the study of ecosystems.

ECOSYSTEMS

Ecosystems are communities of organisms — plants, animals, and protists. The organisms do not live alone. They act upon and are acted upon by the natural substances in their surroundings or *ENVIRONMENT.*

The Organism

An organism is a complete living thing. You are an organism. Within you are the cells that form your tissues and organs. All about you are the living and non-living things with which you interact. You interact with the air when you inhale oxygen and exhale carbon dioxide. You interact with plants when you eat the food and use the oxygen they provide. You interact with all the organisms who share the earth with you.

Populations

Any group of individuals of one species in one area makes up a *POPULATION*. The grizzly bears in Yellowstone National Park are one population. So are the deer in the Shenandoah. The redwood trees in California are another. All of the people in the United States make up its human population. The bass in the pond form a population of interbreeding fish.

Populations are *constantly changing*. As people are born and die, the population of this country changes every day. Robins eat worms and raise new robins, changing both the populations of worms and of robins.

Habitats

The kind of place where a population of organisms is usually found is its *habitat*. Each population is best adapted or suited to live under special conditions. Bullfrogs live on the edges of ponds and lakes, but not in the middle. Tuna fish live only in the salty water of the ocean. Rainbow trout form a population that is adapted for life in a habitat that is a swift-running part of a fresh-water stream.

Communities

Throughout the world are many populations that live in *COMMUNITIES*. Bears and rabbits and fir trees and grass are populations of organisms, but they can all live together in the same forest. This forest is their community.

A *community* is a group of populations of different species living in one place. A farm pond is a community of many species. It may contain cattails, algae, turtles, ducks, and several kinds of small fish. Even the worms wriggling in the soil are part of the community.

The populations in a community *interact* in many ways. The leaves fall from the trees to the forest floor, where some organisms cause decay and return of material to the soil. Squirrels and trees help each other — some of the nuts the squirrels will bury will grow to become new trees. A hawk living in the forest may eat the squirrel and use the tree as a nest. Large trees with thick foliage determine how much light gets through to other plants. *All the members of the community affect each other*.

Ecosystems

An *ecosystem* is the interaction of communities of organisms with their non-living environment. The study of ecosystems is called *ecology*.

A pond ecosystem is a somewhat *closed* system. Bass, sunfish, and perch live among the plants, such as pond lilies, duckweed, and arrowhead. Protists like green algae and diatoms store energy of the sun. A turtle captures a fish, and a water snake eats a frog. An ecosystem is a beehive of activity.

Though water flows in and out, carrying nutrients, oxygen, seeds, and some animals, most of the food eaten by the animals in the pond comes from the plants in the pond. All of the living things in the pond ecosystem require energy to live. The energy enters the ecosystem from the sun. Energy stored in plants goes to small and then larger animals.

In the pond, minerals such as calcium are dissolved by the water, absorbed by the plants for their growth, and transformed into bones of fish, frogs, and turtles when these animals eat the plants or each other. When an animal dies, its body decays, and the minerals are released into the water to be absorbed by plants all over again. Nothing is ever lost or wasted.

Succession

Ecosystems change over periods of time. While ecosystems are not built to last forever, some are less stable than others, and are changed more quickly. Many ecosystems go through a process of change called *succession*, often becoming entirely different.

A pond is an *unstable* ecosystem. Eventually it will fill up with leaves and soil washed in by the stream that feeds it. After it fills up, it will become a *meadow*. Fish and cattails will no longer live there, but mice and grasses will. Soon bushes will shade out the grass, then trees will overgrow the bushes. The unstable pond has changed to a meadow. The meadow, too, will change. Larger plants, and then trees, will take over.

An oak, maple, and hickory forest is a *climax community*. This forest is the last stage in the succession. It may survive for thousands of years. A volcano, forest fire, or an earthquake might destroy large parts of the forest. The stage is set for some steps of succession to repeat.

Ecology

Ecologists are people who study ecosystems. They may study ponds to find out how all of the living and non-living parts work together. They may want to know which types of fish are able to live and reproduce in the pond.

They may add nutrients or plants or snails to see what effect the change will have on the system as a whole. Ecologists may study other ecosystems such as forests, mountains, or deserts.

Pollution in Ecosystems

People produce a variety of waste and waste products that affect the balance of nature in unexpected ways. Sometimes these wastes *pollute* water, air, and soil and interfere with natural succession. Pollution is any undesirable change in an environment. Pollution is most often connected with human activities.

Lakes are often affected by pollution. Topsoil that is washed off from construction sites and improper farming practices fills up streams and lakes. Fertilizer washes off, polluting the water with chemicals that make undesirable plants in the lake grow faster. Industrial wastes dumped into the water kill fish and cause damage. What happens in one ecosystem affects other ecosystems.

The Biomes of the World

The world has many ecosystems. Some are in the ocean. Others are in the desert. There are different climates and surroundings throughout the earth. Organisms live where they can survive, so different plants and animals live in different regions of the earth. These large regions of the earth are called *BIOMES*. There are groups of ecosystems within each biome.

The desert biome, the ocean biome, the grassland biome, and all the biomes on earth together form the *BIOSPHERE*.

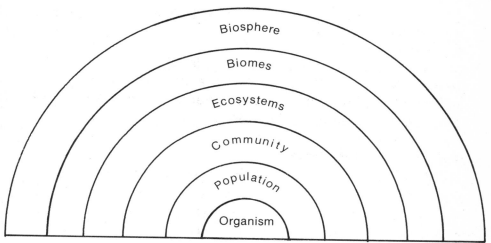

Levels of Organization in the Biosphere

The Biosphere

The biosphere is the part of your planet where life can exist. It includes both living and non-living things. The biosphere is very large compared to you, a single organism. But, compared to the entire planet Earth, it is very small. Think of the world as an orange or an onion. The biosphere is like the peel of the orange or the skin of the onion. It is the thin layer on a large sphere. It is the active part of the sphere.

All substances in the biosphere are *INTERDEPEN-DENT* and need each other to make life work. *All known life exists in the biosphere.* All of the living things are tied to other living things and their surroundings. Organisms interact with each other and with their non-living environment. *Bio* means "life." The biosphere is the area on the surface of the earth where life exists.

When you look at the earth as a whole, the biosphere seems to be very small. The thin layer can easily be damaged. Human beings are the only organisms aware of the biosphere. Humans, in a large sense, are in charge. Humans need to be aware of how the biosphere works. You need to be gentle with your part of it and take care of it. Survival of living things depends on the condition of all of its parts. The whole earth is interrelated. What happens in your part of the biosphere affects your neighbors in their part.

All life on Earth is limited to a thin layer of soil, air, and water around our planet.

CHECK YOUR UNDERSTANDING

1. What is the difference between a population and a community?

2. What effects do trees have on other members of the forest community?

3. What is a habitat? Write down the names of as many kinds as you can think of.

4. Is your backyard an ecosystem? List as many species as you can think of that live there.

5. Describe what happens when a meadow is allowed to go through a natural succession. What does it become?

Key Idea #2:

- Food chains and food webs are about the food and energy relationships in an ecosystem.

FOOD CHAINS AND FOOD WEBS

Links in the Food Chain

Organisms feed on other organisms in an ecosystem. A feeding pattern known as a *FOOD CHAIN* is formed. Big fish eat little fish, and in their turn are eaten by even bigger fish. That describes a part of a simple food chain. The little fish ate water plants to begin with, or maybe fed on insects and insect larvae that fed also on plants, or protists. Actually, all food chains have their start with plants.

Plants as Producers

Green plants and some protists make their own food. They use the energy of sunlight to carry out photosynthesis. Because they do in fact produce food for themselves, they are called *PRODUCERS*. They are producers of food used by other organisms in the ecosystem. Energy is the main player in any food chain, and the producers are the first link in the energy chain. Every food chain begins with some plant producer.

Consumers

Anything that does not contain chlorophyll, including all animals and animal-like protists, must get food from outside their bodies. Since plants and some green protists such as algae are the only food producers, animals must eat them. Organisms that eat other organisms are called *CONSUMERS*. Humans are among the world's greatest consumers. Supermarkets, delis, restaurants, and sidewalk vendors all point to the piles and piles of food of all kinds that humans devour.

Why must consumers eat on a regular basis? Food fills two essential needs: 1) energy to run things, and 2) materials for growth and repair of parts.

Muscle, bone, skin, teeth, hair, and blood are all products of the food you eat and the food you have eaten in the past. You and other consumer organisms must continue to eat to maintain body parts and body functions. The food consumed keeps the body parts in good working order.

Levels of Consumers

Levels of consumers are either first order, second order, or third order. It depends on when they consume and when they are consumed. Eating or feeding is the biggest business of living things. *FIRST-ORDER CONSUMERS* eat plants. Rabbits are first-order consumers, and so are cows. *SECOND-ORDER CONSUMERS* eat other animals. A hawk eats a rabbit. Each level of consumer depends upon the energy stored in the body of the organism that is being taken as food.

Foxes eat rabbits. Foxes are second-order consumers: fox — rabbit — grass.

The Food Pyramid

If you have seen movies of the African plains where lions live, you know that the plains are covered with grass — billions upon billions of grass plants. Herds of antelopes may number in the thousands of animals, but the number is far smaller than the number of grass plants. In the same area, there may only be a few dozen lions. Lions need a wide range in which to hunt.

Any food chain starts with large numbers of producers. The number of consumers falls sharply as you go from one level to the next higher level of the chain. You may think of the food chain as a pyramid with the highest-level consumers at the top. There cannot be very many consumers at the top because it takes so many levels below them to keep them supplied with food. Yet even they are *dependent* on the producers.

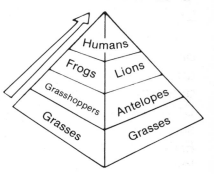

Food Webs

Very few consumers eat only one kind of food. If they did, the failure of the harvest or hunt for that one food would force them to starve. Most consumers eat a variety of foods. The American crow is an example of a bird that eats a wide range of foods.

A frog in a pond is a second-order consumer. The frog eats insects, fish, other frogs, and earthworms. The frog in turn is eaten by fish, snakes, and some birds. Frogs are part of a *FOOD WEB* like most other consumers. A food web is a complex pattern of overlapping food chains. It shows the feeding *interactions* in an ecosystem. It shows which animals depend on which other animals for energy and for materials from their food. A food web shows that organisms depend on the producers in the community. A food chain is only a part of a food web, showing one line of the relationships or connections. Both chains and webs show how energy and materials travel from one member of the community to another.

Decomposers

Suppose that a high-level consumer like the lion lives to a ripe old age and dies in a place where it is not eaten by a *scavenger*. Does the food chain stop there? No, that is where the *DECOMPOSERS* enter the scene.

Bacteria are everywhere, and so are molds and other fungi. Whenever an organism dies, if nothing eats it directly, it will soon be broken down, digested, and taken in by these decomposers. The job of the decomposers is to turn the food chain into a *FOOD CYCLE*. Decomposers break down the *complex materials* of a dead animal or plant into simpler raw materials that can again be taken in by plants (producers) to make more food. Nothing is ever wasted.

CHECK YOUR UNDERSTANDING

1. What is the difference between a food chain and a food web?
2. Why can plants get along without animals, but animals cannot live without plants?
3. Diagram the longest food chain you can think of in the form of a pyramid. How common is the animal at the top compared to the bottom organisms?
4. How do decomposers work?
5. What is a food cycle?

Key Idea #3:

- All substances are constantly reused in a never-ending series of cycles as living things carry out basic life activities.

SPACESHIP EARTH

The planet Earth has been compared to a spaceship — a very large one, but a spaceship nonetheless. The earth is a planet, and we inhabit its outer surface. We use the materials we find here to build our homes, to make tools, and even to eat. If some substance is already in scarce supply, there is no way to get more of it. There are no cargo spaceships bringing gold from other planets, no trash spaceships carrying away waste. Everything that is here now, stays here. The biosphere is limited to what is here now. Like a spaceship, the earth is isolated in space.

The earth is a *closed system*. Only energy from the sun enters, and only heat energy from the earth is removed.

Everything else on the earth is reused. Water to drink is the same water that was here millions of years ago. It is used, cleaned up, and used again. It is cycled and recycled. When you drink water, you may be using some water that was once drunk by George Washington, or a dinosaur, and maybe even yourself, years ago.

The Water Cycle

Constant reusing of water is called the *water cycle*. Water from puddles, oceans, lakes, and rivers evaporates and becomes a gas. When a puddle "dries up," the water evaporates and becomes part of the air. Water vapor comes from other places, too. The leaves of trees, the soil, and even your breath are sources of water vapor. The oceans are the largest source of water that evaporates into the air. Oceans are big contributors to the water cycle.

There is always water vapor in the air, but you cannot see it. Even though water vapor is invisible, you can tell that it is there. A glass with ice in it is cold, and water vapor *condenses*, or becomes liquid, on the outside of it. The water that forms on the outside of the glass comes from the air, not from inside the glass.

Warm air holds more water vapor than cold air. Warm air rises, and when it does, it gets colder. The higher parts of the atmosphere are colder than the surface air. When the temperature gets lower, the air can hold less and less moisture. The water vapor *CONDENSES* and becomes water droplets. As temperatures change to colder and colder, the droplets gather closer, get bigger and heavier, and fall as rain when conditions are right. Water can be used over and over again by plants, animals, protists, and people.

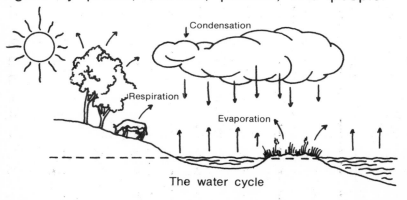

The water cycle

The Carbon Cycle

Carbon is one of the most abundant and useful chemical elements in the world of living things. Carbon is a big part of our bodies and of almost everything we eat. All living material is made up of chemical compounds made with carbon. Except for water, all of the basic life chemicals contain carbon.

There would be no proteins, fats, or carbohydrates without carbon. All food except water and salt contains carbon.

Green plants make food using the carbon dioxide gas in the air. Protists, largely in the oceans, make a great deal of food. Animals make carbon dioxide as a waste product and put it back into the air when they breathe. Together, these producers and consumers constantly *recycle* carbon, using it over and over again.

Some carbon gets tied up in such things as fossils and fossil fuels like coal and oil. Eventually, it gets back into the cycle to be used and reused.

The Oxygen Cycle

Oxygen is one gas that is essential to almost every form of life. Plants, animals, and protists use oxygen to change their food into the energy they need to survive. Even the producers in the food chain need oxygen in order to convert the food they make into energy to maintain their own life.

All of the oxygen in the air comes from producers. Green land plants and ocean protists provide oxygen for all living things. When the sun is shining, green plants not only recycle the oxygen which they use, but are able to release more than enough of the vital gas for the consumer organisms.

The Nitrogen Cycle

Nitrogen is a gas that makes up about 78% of the air. When combined into chemical compounds, it is absolutely necessary for life. Many important compounds, especially proteins, contain nitrogen.

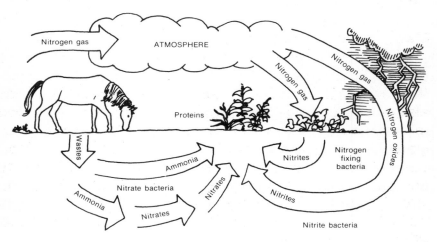

The Nitrogen Cycle

The nitrogen in the air is not usable for making proteins directly. It must first be combined into a chemical compound such as ammonia. The job is done by bacteria in the soil which are called *nitrifiers*. They take nitrogen from the air and change it to a form called *NITRATES*. The nitrates in the soil are used to make fertilizer for plants. The plants absorb the nitrogen compounds and make proteins which consumers can eat.

Consumers produce waste which contains other nitrogen compounds that plants can use. Once nitrogen is taken from the air by nitrifiers, it can go from plants to animals and back again in different forms. Cattle eat plants to get protein, and their manure makes an excellent fertilizer for the plants.

There are other bacteria in the soil that undo the work of different nitrifiers. They are called *denitrifiers*. They break down proteins and other nitrogen compounds and return the nitrogen to the air. The amount of nitrogen in the biosphere, like other elements, stays the same: cycled and recycled, but never lost.

Cycles Work Together

None of these cycles works alone. Although a cycle can be described for each element or compound, they are really joined in nature. Nitrogen compounds also contain carbon and oxygen, and they are dissolved in the water of the water cycle. The whole ecosystem works together like a big, complicated machine. You can look at one part at a time to make it easier to study and understand. Each cycle is only a small part of the whole system. Systems are made up of working parts that do things in cycles. The cycles interact with each other.

Every substance in the ecosystem is recycled over and over again. There is a cycle for iron, calcium, phosphorus, potassium, and every chemical used by living organisms. Biology is the *study of life*, but is dependent on the recycling of the *non-living chemicals* in the biosphere.

CHECK YOUR UNDERSTANDING

1. What is a closed system?

2. What makes water vapor condense high in the air?

3. Why is carbon important to living organisms?

4. What organisms take nitrogen directly from the air?

5. Why does nature recycle everything? What would happen if something were not recycled?

> **Key Idea #4:**
>
> - Energy is essential to all life. Energy constantly enters and leaves the earth and cannot be recycled.

WHAT IS ENERGY?

Energy comes in many forms. Light and heat from the sun are energy. Electricity is also energy. Your muscles need energy to run and throw a ball. Your heart uses a lot of energy to pump blood, and your brain uses it when you think. The warmth of your body is heat energy.

Food Is Energy

Energy can be stored in chemicals like gasoline, to be released when it burns. Food is another form of stored energy.

Producers absorb energy from the sun when light falls on leaves that contain *chlorophyll*. Only part of the sunlight is absorbed, and only part of that is stored as food. Some of the food is used by the producer itself. Only a small part can be passed on to consumers.

Consumers Use Energy

The way in which consumers use energy is like the levels of a pyramid. The first or lowest level of the pyramid is made up of producers like grass or plants. The energy stored in the plants is the greatest amount of energy at any level. The next level of organisms in the pyramid are the first-order consumers, or plant eaters. Rabbits are a good example. First-order consumers like rabbits eat producers like plants and protists. The rabbits receive much, but not all, of the energy held in the plant. Plants use some of it to keep going. The next level of the pyramid is made up of second-order consumers or animal eaters. Foxes are an example. Foxes capture and eat rabbits. The foxes get a good deal of energy from the rabbits, but not all of it.

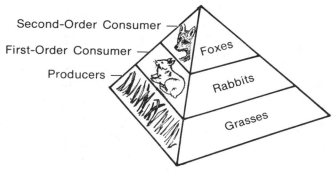

Second-Order Consumer —

First-Order Consumer —

Producers —

Foxes

Rabbits

Grasses

An Energy Pyramid

Consumers do not produce any energy of their own. They must get all their energy from producers. It takes many producers to keep one consumer alive. A few rabbits must eat a field full of plants to get enough energy. In the same way, it takes a number of rabbits (first-order consumers) to keep just a few foxes (second-order consumers) alive. There is a loss of energy at each step of the food chain. Less of the original amount of energy is available the further away the consumer gets from the producer.

Like a pyramid, there are a great many plants at the bottom, a small number of rabbits in the middle, and just one or two foxes at the top.

Second-order consumers, like foxes, eat first-order consumers, like rabbits, to get their own food energy. However, only a small part of the food energy eaten by the rabbits is left to be passed on to the foxes. The loss of heat energy at each step of the food chain explains why there must be large amounts of energy to start with. The larger number of plants as producer organisms is shown in the form of a pyramid. Less and less of the original amount of energy from the grasses is available to each new level of consumer.

Heat

A warm-blooded organism constantly loses heat to the air which surrounds it. Waste energy ends up as heat. At each step of the food chain, less energy is stored. It becomes heat. Finally, all of the stored energy becomes heat, and there is no energy left to be recycled.

Energy cannot be recycled. And that is why the sun is so important every day. The sun sends energy back into the system.

Energy from the Sun

Producers are dependent upon a constant supply of energy in the form of sunlight. If the supply were to stop, no food would be produced, and in a short time, all the stored food energy would be used up.

Energy comes to earth from the sun every day, even when it is cloudy. Light and heat are absorbed on the daylight side of the earth constantly. The same amount of waste heat energy is constantly lost to space on the night side at the same time. The earth constantly takes in and loses energy at a rate that keeps it from getting hotter and hotter overall. Each spot on the earth goes through warm periods and cool periods daily and yearly, but on the average, the temperature of the whole earth remains the same.

You cannot make your own food from sunlight as the producers can. All of your food comes from the sun indirectly. You and all living organisms depend on the sun for life itself.

CHECK YOUR UNDERSTANDING

1. Why do producers need sunlight?

2. Why doesn't the earth get hotter or colder?

3. Moles live underground and eat insects. Why do they need the sun?

4. Where can energy be stored?

5. What is the difference between a producer organism and a consumer organism?

6. What does a third-order consumer do?

SUMMARY OF CHAPTER 12

Every living organism depends on other organisms for survival. The non-living parts of the environment are equally important. Together the interaction of living and non-living things make up an ecosystem.

All food comes from plant and protist producers directly or indirectly. Food energy is really energy from the sun stored by the chlorophyll in the producers. Both plants and animals depend on sunlight and chlorophyll.

All substances on earth are constantly reused in a never-ending series of cycles. Some especially important cycles are those of carbon, oxygen, nitrogen, and water. All cycles work together to keep the biosphere running.

Energy is the only thing that constantly enters and leaves the closed system of the earth. It is necessary for life and cannot be recycled.

WORDS TO USE

1.	pollutant	19.	food chain
2.	succession	20.	first-order
3.	ecology	21.	second-order
4.	environment	22.	dependent
5.	biosphere	23.	water cycle
6.	climax community	24.	condense
7.	pollution	25.	recycle
8.	population	26.	oxygen
9.	closed system	27.	carbon
10.	ecosystem	28.	evaporate
11.	habitat	29.	atmosphere
12.	interact	30.	cycle
13.	community	31.	nitrifier
14.	biome	32.	sunlight
15.	food web	33.	denitrifier
16.	decomposer	34.	energy
17.	producer	35.	nitrogen
18.	consumer	36.	chlorophyll

• REVIEW QUESTIONS FOR CHAPTER 12 •

1. What is the biosphere?
2. Why do ecologists study pieces of an ecosystem instead of the whole thing at once?
3. Is the earth an open or closed system? Does anything enter or leave it?
4. What does every food chain start with?
5. What is the difference between the amounts of energy lost from each level of the food chain?
6. What does the word *photosynthesis* mean?
7. What is an environment?
8. Why is a pond community not permanent?
9. Why do many species live only in certain habitats?
10. What is ecology?
11. What is the last stage of a succession called?
12. Name three ways people pollute their environment.
13. What do producers produce?
14. Which could survive better alone — producers or consumers? Why?
15. What two needs are filled by food?
16. Why is a food chain like a pyramid?
17. Why is the earth like a spaceship?
18. How old is the water you drink?
19. What substance is found in all of the basic life chemicals except water?
20. What do producers provide for humans in addition to food?
21. How is nitrogen from the air made into proteins?
22. Why don't we run out of water, oxygen, or any other substances in nature?
23. Why do we need a constant supply of energy?

Chapter 13

Describing the Behavior of Organisms

Chapter Goals:

To recognize that patterns of behavior are the responses of organisms to conditions in the environment.

To understand that the survival system in populations of organisms includes innate, learned, and communication behaviors.

Key Ideas:

- Certain behaviors of each species are innate, or inborn.

- Learned or acquired behavior involves the use of thinking, reasoning, and problem solving.

- Communication is a form of behavior that has survival value for populations of organisms.

INTRODUCTION

BEHAVIOR is a term used to describe all of the things that organisms do. Behavior is mainly of two kinds: 1) those things that are done automatically and are *INNATE*, and 2) those things that are *LEARNED* or *ACQUIRED*.

Innate behavior is present at birth. It does not have to be learned. Heredity is responsible for built-in pathways that cause organisms to behave automatically in certain ways. Learned behavior is behavior that occurs after an animal has had some life experiences. Learned behavior is a characteristic of all vertebrate animals and is most highly practiced in humans.

Humans study behavior of animals because they are curious and because they see patterns of their own behavior in the behavior of other organisms.

Key Idea #1:

- Certain behaviors of each species are innate, or inborn.

SPECIES BEHAVIOR

Innate Behavior

INNATE BEHAVIOR is controlled by genes that are inherited. Humans inherit a *REFLEX* that causes blinking when any object moves near the eyes. From the time of birth and all through life, the blinking reflex protects the eyes from harmful objects. Blinking occurs without thinking. It is a reflex act. The nerve pattern for blinking is inborn. The response is automatic.

Nest-Building Behavior

Nest building is an inborn set of behaviors. Nest building is an *INSTINCT*. Birds build a wide variety of nests. They are built in different places, with different materials, and they come in all sizes. Materials for nests are a part of the *HABITAT*. A habitat is that special part of an ecosystem where *EACH SPECIES* lives its particular lifestyle. *Gulls* use seaweed and other seashore plant materials to build a large nest on the ground. *Terns* living in the same community make a circular dent in the sand where they lay their eggs. These nest-building behaviors are innate or inborn. Each kind of bird builds one kind of nest. And they build it in a certain place in their habitat.

Courtship Behavior

Nest building is just one part of a set of behaviors that are all connected and all inborn or innate. *COURT-SHIP PATTERNS* come ahead of nest-building and mating behaviors. Courtship is the set of behaviors that are used in attracting and acquiring a mate. Courtship involves a pair of organisms, but the male animal does most of the courtship display. All of the vertebrate animals exhibit courtship behaviors.

One part of the courting process is making some kind of sound. In birds it involves the song of the male. In frogs it involves the croaking of the species mating call. In a pond where several species of frogs are courting, it is important for each species to mate with its own kind. Survival of the species depends on having the right sperms with the right eggs. Otherwise proper development does not take place, and the members of the next generation are lost.

Behavior and Survival

Courting behaviors are inborn or innate. The life span of some of the lower vertebrates is not long enough for learning all of the steps involved. Genetics provides the hereditary materials that have the chemical and nervous signals all lined up. Courting behavior leads to mating. Much of this behavior must be timed perfectly. There is no room for *TRIAL AND ERROR.* The survival of the species is built into a heredity pattern that causes and coordinates the behaviors involved in courtship, mating, and nest building.

Incubating the eggs is a part of bird instinct. Sometimes both birds share the incubation task, while other times it is the job of one or the other of the parents.

Territorial Behavior

A *TERRITORY* is an area that is staked out by a species of animal. It usually includes a nesting place and the surrounding ground that provides the main source of food for the species. Birds have territories that they claim by their song and by fending off intruders. The mockingbird in your neighborhood finds the tallest of trees to perch atop and sing its *TERRITORIAL SONG.* In neighboring territories other male mockingbirds sing to claim their own space.

The advantages of a territory to a species are tied to the whole life of the organism. Having an established territory is useful in getting a mating pair together. It allows them to build a home or nest and raise their young. The borders of the territory are defended, and most intruders are driven away. Fighting over the territory is reduced by the fact that other organisms usually honor the "owner's" rights. Animals that are successful in establishing a territory do so because it is their song that attracts a mate, or they are able to *DOMINATE* in the fights to protect their area. Dominant behavior establishes the leader or the winner.

SURVIVAL of the species is tied to the ability of a population of organisms to reproduce. Having a territory promotes the process of courtship, mating, nest building, rearing of the young, and reproduction. The next *GENERATION* that results from successful parent organisms has a genetic makeup that favors future success. Territorial behavior insures a good food supply for a family, an uninterrupted breeding pattern, and reduction of fighting and killing.

Imprinting Behavior

Newly hatched geese are called *GOSLINGS*. Ordinarily they attach themselves to the mother goose. In experiments to study behavior, the eggs of geese were removed from the nest and hatching took place in front of the experimenter. The baby goslings attached themselves to the first object they saw. In one case the experimenter was Konrad Lorenz, a famous biologist who studied behavior of animals. He described this social behavior of the goslings and called it *IMPRINTING*. Goslings follow their "first sight" leader whether it is their natural mother or whether it is a person or even an object. Lorenz found out that once imprinting behavior gets set, it does not change. Geese followed him everywhere. He became their "parent."

Imprinting is "instant learning" and is of *survival value* to the species. Goslings in the wild become a fixed part of the family unit. This behavior plays an important part in the life cycle. By keeping the goslings in the family circle, it encourages courtship, nest building, mating, and rearing of the young.

Goslings attached to a hen instead of their own mother.

Spider and Bee Instincts

Instinctive behavior is not limited to birds. Spiders make and use *silk* for building their webs and lining their nests. *Weaving a web* is a complex set of behaviors that each species of spider inherits. Newly-hatched spiders do not get a lot of time with their parents. They are on their own so quickly that they could not survive if they had to go through a long web-building learning process. How to build a web is important in capturing food. Wrapping a captured insect in silk would be difficult to learn. So web-building and insect-wrapping depend entirely on the instincts of the spider. Without food, no spiders!

The same is true for *bees*. The life of the bee is one continuous set of instincts. Building the hive, finding and getting food, caring for the queen bee, protecting the hive, and swarming, are all inborn behaviors. There is no time in the life span of the bees for learning how to do all of these things.

Bees build
six-sided
cells.

Plant "Behavior"

Behavior is usually thought of only in connection with animals. Behaviors are observable changes. Plants do make *some* responses and changes to stimuli. Light, dark, gravity, touch, and chemicals cause "behavior" in plants. Parts of plants turn toward the light, roots grow toward the center of gravity, and flowers and leaves open and close.

Phototropism. Potted *geraniums* placed on the window sill will show the broad part of the leaves facing the source of sunlight. When the plants are turned around so that the broad side of the leaves face the room, it takes just hours before the plants turn back around. The sunlight causes chemical changes that make cells on one side of the plant get longer. The plant is forced to bend toward the light as the longer cells on the outside act against the shorter cells on the inside. The tug-of-war turns into a push of war. The longer cells are successful!

Daylight has an effect on the opening of *morning glories*. The flowers open to the fullest extent, and then, as evening approaches, the flowers curl closed for the night. The reaction of plants to light is called *PHOTO-TROPISM*.

Geotropism in Plants. Radish seeds and bean seeds planted in the laboratory showed that roots grow "down." Growing "down" is a response to gravity and the behavior is called *GEOTROPISM*. What would be your guess about which way the *stem* of plants would grow?

Response to Touch. The closing of a *Venus flytrap* is in response to the presence of an insect landing on the sensitive "hairs" that line the outside of the leaves. The pressure of the insect on the fine "hair" causes pressure changes inside the plant, and it closes and captures the insect.

Another touch response occurs in the *mimosa*, or "sensitive plant." Rows of leaflets, usually extended, are brought together by the pressure of touch. The closing happens in a matter of seconds. The twining of vines around tree trunks, fences, posts, and trellises is a "behavior" in response to the touch of the object being wrapped around.

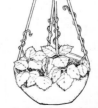

A twining vine

A major feature of cells of plants and animals is that they are made of material that is *SENSITIVE* to changes in their surroundings. Individual cells like ameba and paramecium react to chemicals and objects in the environment. The stuff that cells are made of is often called *PROTOPLASM* or *CYTOPLASM*. This material is unique among all of the materials of the world. One of its main features is that it reacts or responds to stimuli. It is in this active and responsive material that the foundation or basis for the life activities is located. You have probably heard the word *SENSITIVE* used in relation to a "sensitive person" — a person who is responsive to the actions of other people. Cells and tissues of animal and plant bodies are sensitive in that they respond to signals from objects and materials in the environment.

CHECK YOUR UNDERSTANDING

1. What is a main feature of an innate behavior?

2. What makes blinking an innate behavior?

3. What is the best way to describe nest-building behavior in birds?

4. What is the purpose of courtship behavior?

5. What is the value of territorial behavior in animals?

6. What is imprinting behavior?

7. What is the value of instincts in spiders and bees?

8. What is similar about animal and plant behavior?

9. What is the difference between phototropism and geotropism?

10. What does it mean to be sensitive?

> **Key Idea #2:**
>
> • Learned or acquired behavior involves the use of thinking, reasoning, and problem solving.

LEARNED BEHAVIOR

Conditioned Response

An aquarium full of goldfish or guppies has a regular feeding time of 5:00 P.M. As the keeper approaches the aquarium at 5:00 P.M., the fish move to the top and to one corner of the aquarium. The keeper taps on the rim with the container holding the fishfood. The fish scurry about at the surface. The *BEHAVIOR* of the fish is a type of *LEARNED BEHAVIOR*. Their movement to the surface is a *RESPONSE* to the *STIMULI* of the 5:00 P.M. approach of the keeper and to the tapping sound of the food container. The food itself is the *PRIMARY STIMULUS*. What do you think would happen at other times during the day if the keeper simply walked over to observe the fish? If you said that they would just go about their business as usual, you would be right! The *BEHAVIOR PATTERN* of the fish is part of a learned behavior called *CONDITIONED RESPONSE*. The fish have become conditioned to the 5:00 P.M. feeding time, the approach of their keeper, and to the tapping sound of the container followed by the presence of food.

Pavlov's Dogs

Learning about what behavior is has taken some complicated behavior on the part of scientists. Ivan Pavlov, a Russian scientist, was interested in studying the digestive process and feeding behavior of dogs. His work describes the kind of learning known as *CONDITIONED BEHAVIOR.* Production of saliva in the mouth is a *REFLEX* action. Dogs salivate a great deal. They produce saliva at the sight and smell of meat, their preferred food. Dogs swallow *whole pieces of meat*, and the flow of saliva helps that happen. Saliva helps the food get down the gullet to the stomach.

Pavlov figured out a way to collect the saliva and measure the amount of flow. He showed the dogs meat, and collected the saliva. When the dogs were accustomed to the practice, Pavlov began ringing a buzzer at the time he presented the meat to the dog. After a number of repeated buzzer and food servings at the same time, Pavlov then rang the buzzer by itself. *The dogs produced saliva in amounts the same as when they received meat.* The sound of the buzzer alone caused the dog to show the same behavior as the natural sight of meat.

Pavlov found that if he continued to press the buzzer without offering the food at the same time, the dogs' flow of saliva was reduced. After repeated use of the buzzer with no food, the saliva flow was reduced to nothing. The dogs had learned to associate the buzzer with food. The natural flow of saliva at the sight of good food is *INNATE BEHAVIOR.* Response to the buzzer and food was a *CONDITIONED RESPONSE.* Conditioned responses are a learned behavior. Learned behaviors can be "un-learned." So it was with "the buzzer and no food" response. Pavlov earned the Nobel Prize for his work on digestion.

Skinner's Rats

B. F. Skinner, an American scientist, used the rat, *Rattus rattus*, as the subject of his experiments on behavior. It turns out that rats are always hungry. Skinner designed a box in which a rat could press a lever, and a food pellet would appear. The rat, following its pattern of always being hungry, continued to press the lever. Every time the lever was pressed, it was automatically recorded. Skinner found this went on for hour after hour. Just as in Pavlov's dogs, if the food supply is cut off, the rat stops pressing the lever after a while.

Skinner went a step further by giving food when the lever was pressed, but *only when a light was on* in the box. In the dark, the rat did not press the lever bar. *The rat learned the difference between "food in the light" and "no food in the dark."* The rat had figured some things out. The rat found out that its behavior caused something to happen. And furthermore, the rat learned that the light made a difference.

Adaptation

The key difference between the dog experiment and the rat experiment is that the dog response was a *REFLEX* with saliva flowing just before the food was taken. In the rat experiment, the food appeared *after* the lever was pressed. Learning that food would appear with the pressing of a lever led to success in food-getting for the rat. The rat had *adapted* to a set of conditions in the environment. Adaptations in natural populations of animals lead to similar successes. Successful adaptations are helpful to the continuation of the species. Species are able to continue if a population can adapt to change in the environment.

Bird Song Behavior

Is the song of the mockingbird or the chaffinch a *learned behavior* — or is it inborn? Research with birds raised in *ISOLATION* indicates that a basic bird-song pattern is inherited. Birds hatched from eggs taken from the nest and brought into the lab are able to sing the species song. But, they do not sing the *elaborate*, or full, song of the parent birds. When placed in the presence of the parent birds, there is some learning that takes place by *imitation*. Birds and *PRIMATES* have been the major animals used in the study of behavior.

Reasoning

The key to *LEARNED BEHAVIOR* is that it depends on the ability to reason. *REASONING* involves making connections between ideas. Learning involves memory, which is the ability to recall or reproduce ideas. How memory works is not completely understood. Reasoning involves the use of *LOGICAL PATTERNS* of thinking. The end result of learned behavior is the ability to solve problems.

Chimpanzee Behavior

A biologist placed a chimpanzee in a room by itself to find out what it might do. When the scientist peeped through the keyhole, what do you think he saw? Yes, the eye of the chimpanzee was looking out of the keyhole from the other side. The chimp was just as curious as the scientist!

The same scientist put a banana on a string and hung it from the ceiling. The banana was out of reach of the chimpanzee. Several boxes were placed on the floor around the room. When the scientist looked through an observation opening in the door, he saw that the chimpanzee had the banana. The chimpanzee had

solved the problem by putting the boxes on top of each other, climbing to the top, and easily reaching the banana. The chimp had put some ideas together. The problem involved use of the brain. Reasoning took place.

Problem Solving

PROBLEM SOLVING is the highest kind of learned behavior. Living things are limited in the kinds of behavior that they can use by the kind of nervous system they have. Learning behaviors are features of higher level vertebrates. The ability to learn and to solve problems governs the lifestyle and patterns of activity of organisms. Human beings have the *most complex behavior patterns* of all living things. Humans are the best problem solvers. They are the best at solving problems because they have developed the highest system of *COMMUNICATION*. Complex and useful *LANGUAGE PATTERNS* separate humans from all other organisms.

CHECK YOUR UNDERSTANDING

1. What is the name given to the behavior of fish in an aquarium when they come to the surface for food?
2. What reflex action was involved in Pavlov's dogs?
3. What part did the buzzer play in Pavlov's work?
4. What conditioned response did Pavlov discover?
5. What is a major feature of learned behavior?
6. What was learned about the song of birds by raising them in isolation?
7. What is a feature of the behavior of chimpanzees?
8. What is the highest kind of learned behavior?
9. What separates humans from all other organisms?

```
Key Idea #3:

• Communication is a form of behavior
  that has survival value for populations
  of organisms.
```

COMMUNICATION BEHAVIOR

What Is Communication?

COMMUNICATION is an activity of an organism that involves the use of some form of signal or stimulus. Communication is behavior. Communication may be a signal from an organism to another of its kind or to other kinds of organisms in the environment. Most communication behaviors take place between members of the same species.

Bees have a system of communication related to finding food. They do a dance pattern that indicates to other bees the direction and distance of flowers in their environment. Knowing where to find food is important to the survival of the hive and the species.

Birds have elaborate communication systems for getting and keeping a mate. They communicate through song to keep their own territory. Baby birds open their mouths to be fed as the beak of the parent bird communicates by markings. Male birds get into fighting postures with other males to keep their dominant place in the flock. Most birds also have a communication signal that causes the group to take flight at some sign of danger to the flock.

Sign Language

Humans have tried to teach chimpanzees and other apes how to use the *sign language* of the deaf. Language is a part of behavior. Animals respond to the stimulus of the signs and sounds of communications behaviors. A recent news article told of a chimpanzee that uses "hundreds of sign language words" teaching a baby chimp. The baby chimp learned 55 sign words.

Chimpanzees Teach Sign Language

Scientists Told That Apes in Experiments Instruct Each Other

Also reported was a story of an orangutan that had used sign language to "ask to be taken for a car ride," "took along money it had earned for keeping its room clean," and "gave the driver directions to the local Dairy Queen."

Primates have the highest order of behavior among all of the animals. Among the primates, human beings have the highest level of intelligent behavior.

Human Behavior

The key idea in *HUMAN BEHAVIOR* is that human beings are *REASONING* organisms. The ability to think about everything and to remember an almost endless number of words, ideas, and events is remarkable behavior. Humans are problem solvers, judgment makers, and decision makers.

Humans establish governments and religions, create states, and make laws. Humans run on the land, swim in the seas, fly through the skies, travel in space, wander on the ocean floors, climb the highest mountains, and make their homes in every kind of place. Humans build and run the largest factories and the largest cities.

Humans also pollute the air, water, and land. They foul the atmosphere with hydrocarbons from their automobiles. They have altered the atmosphere in ways that affect *GLOBAL* temperature conditions. This alteration can, over long periods of time, affect the vegetation, climate, and living conditions of all species. One famous writer claimed that "humans are the only animals that blush — and the only animals that need to!"

Fortunately the human species is a problem-solving and decision-making organism. Problems that have been created are able to be solved. Modern technology and communications systems have improved the possibilities to do so.

Human Culture

All of these behaviors are based on the human ability to make and use language to communicate. Because of the written and spoken word, the *COLLEC-TIVE HUMAN CULTURE* — all of knowledge and history — is able to be passed from generation to generation. New populations benefit by having a storehouse of information and ways of doing things that have already worked. New knowledge is added with each generation.

This *CULTURAL EVOLUTION* is strictly a human event. *Homo sapiens* is now a part of both the process of biological evolution and cultural evolution. Changes in human culture are happening faster than changes in biological evolution. Humans are more in charge of events that affect the human species. Humans alter the environment to improve their chances for survival.

Advantages of a Long "Growing-Up" Period

Humans have the longest growing-up period of any organism. Humans are *infants* and *children* for years. They have an *adolescent* "growing up" period and even a long *young adult* period when they have a chance to learn what is needed to get along in a complex world.

Human communications and other kinds of *LEARNED BEHAVIORS* continue all through the life span. Human beings solve problems, gain new knowledge, add it to prior knowledge, and use it to solve new problems. Human behavior is affected by previous behavior. Previous behavior is known as *EXPERIENCE*. Humans often say that "experience is the great teacher" — and so it is!

CHECK YOUR UNDERSTANDING

1. What do bees do to communicate with other bees?

2. Why do birds communicate with each other?

3. In what way have chimpanzees and orangutans been taught to communicate like human beings?

4. What are the ways in which human behavior is different from the behavior of other animals?

5. What things have humans done that are good for this planet?

6. What things have humans done that are bad for this planet?

7. What special behaviors do humans have that enable them to pass their culture down from one generation to the next?

8. What is the advantage of a long growing-up period for humans?

SUMMARY OF CHAPTER 13

The two main types of behavior are innate behavior and learned behavior. Innate behavior is inborn or instinctive. Heredity is responsible for this behavior. An organism is born with this type of behavior. Nest-building, courtship patterns, and displays are examples of innate behaviors in birds. Weaving a web is an instinctive behavior of spiders. Animals are born with imprinting behavior. Even plants instinctively respond to stimuli. Innate behavior helps a species to survive.

Higher organisms, especially vertebrates, can also learn things. Learned behavior is acquired in different ways, such as by conditioned response and by trial and error. The highest form of learned behavior is problem solving. Humans are the best problem-solvers.

All animals have a system of communication. Bees communicate by dancing. Birds use song. But humans can communicate through speaking, reading, and writing. In this way, people can store information and pass it down from one generation to the next.

WORDS TO USE

1. behavior
2. innate
3. reflex
4. courtship behavior
5. displays
6. territorial behavior
7. dominant behavior
8. imprinting
9. goslings
10. instinct
11. learned behavior
12. stimulus
13. response
14. conditioned response
15. Pavlov
16. Skinner
17. isolation
18. logical patterns
19. problem solving
20. communication
21. sign language
22. reasoning
23. experience
24. cultural evolution
25. human culture

• REVIEW QUESTIONS FOR CHAPTER 13 •

1. What is an instinct?
2. What is a habitat?
3. What is the reason for courtship behavior in birds?
4. What is the advantage of courtship behavior being inborn instead of learned?
5. What is territorial behavior?
6. What do courting behaviors, signals, and territorial behaviors do to help a species to survive?
7. What is the value of imprinting?
8. What innate behavior does a spider have?
9. What is phototropism?
10. What is a sensitive plant?
11. What is the difference between innate and learned behavior?
12. What is a conditioned response?
13. What did Pavlov do to condition his dogs?
14. What was the difference between the behavior of the animals in Pavlov's dog experiments and in Skinner's rat experiments?
15. What is the highest kind of learned behavior?
16. What are some methods of animal communication?
17. What are some methods of human communication?
18. In what ways do humans endanger the planet Earth?
19. What is the cultural evolution of humans?
20. What is there about communication that helps people to survive?

Chapter 14

Tracking Changes In Organisms Through Time: Evolution

Chapter Goals:

To trace the fossil record as a means to understanding the history of life on earth.

To describe the main ideas of the theory of evolution and natural selection.

Key Ideas:

- Fossils provide evidence for the existence of different kinds of organisms over long periods of time.

- The theory of evolution and natural selection offers explanations for changes in natural populations of living things.

- Since its early beginnings in the fossil record, the human species has become the most widespread and the most numerous organism on earth.

> **Key Idea #1:**
>
> - Fossils tell the story of changes in organisms on earth.

FOSSIL EVIDENCE

What Are Fossils?

FOSSILS are the remains, impressions, or other traces of animals and plants of the past. Fossils are *PRESERVED* in the earth's crust. They are records of animals and plants of the past. The word "fossil" comes from the Latin word *fossilis*, which means "dug up." Different layers of *SEDIMENTARY ROCK* were laid down at different times. Different kinds of fossils are discovered in the layers of rock. Fossils are uncovered as scientists dig.

Men and women *PALEONTOLOGISTS* and *AN-THROPOLOGISTS* dig for and study fossils. You might wonder why. Humans are very curious about life in the past. The history of life on earth is a long one; that history is called the *FOSSIL RECORD*.

The Fossil Record

The 3-billion-year history of life on earth shows that fossils have been found in almost every part of the world. Any trace of plants or animals from the past are considered fossils. *FOSSIL FUELS* in the form of coal, oil, and natural gas are the products formed from once living things. Other fossils have been found enclosed in *AMBER*. Amber is a yellowish mineral that is transparent. This kind of fossil is called a *MINERALIZED FOSSIL*. Tracks and footprints found in rock are fossils. Another fossil record is in the form of *PETRIFIED WOOD*. Over long periods, chemical changes have caused the wood to change to stone. It is hard, rigid, and heavy. But the original form and structure of the parts of the woody tree have been PRESERVED — a history of plants made in rock.

Another interesting fossil record is in preserved animal DUNG, or solid waste. This fossil record is of special value because it tells the story of types of food eaten by the animals of the past. The story of plant life follows along with the story of animals.

Fossils and Evolution

Fossils provide information about the structure of animals and plants from all parts of the earth. The FOSSIL EVIDENCE gives indirect support to the theory that there have been changes in animals and plants over very long periods of time. Fossils found in layers of rock next to each other are more alike than fossils from layers far apart.

The horse of modern times has been traced in the fossil record through 60 million years. The teeth of horses and their hooves show a series of changes that indicate ADAPTATIONS to changing conditions. Even the condition of the surfaces of the teeth show different kinds of wear and tear. The horse of 60 million years ago was much smaller than modern-day horses.

Fossil Variation

The fossil record shows sea life fossils from as long as 500,000,000 years ago. In the earliest fossil records there was not as much variety of life forms as in the more recent records. The records indicate an increasing number of different kinds of life with the passing of time. Different major groups of animals show up in the rock formations. An "age of fishes" was followed by an "age of amphibians," where each form of life was the DOMINANT one.

The age of the earth has been estimated to be 4.5 billion years. *GEOLOGICAL TIMETABLES* show the first fish fossils at about 500 million years ago. Amphibians appeared in the record about 350 million years ago. Reptiles, insects, and ferns were present about 200 million years ago.

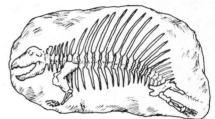

Dating Rocks and Fossils

Finding out the age of a certain rock formation takes some detective work. If you know something that is a fact, then you can build other ideas around what you know. Scientists know that certain minerals are *RADIOACTIVE*: the mineral gives off rays, and it changes to another substance at the same time. *Uranium* is a radioactive element that gives off rays and changes to *lead*. The change is very slow, but it is also very regular. In fact, the *RATE OF CHANGE* of uranium to lead is like a clock ticking off the seconds, minutes, and hours. But instead of the seconds and minutes, the "radioactive clock" has ticked off and measured years.

Uranium has a *HALF-LIFE* of 4.5 billion years. The half-life is the amount of time that it takes for a chunk of uranium to become one-half lead and one-half uranium. The rate of change stays the same. Because the rate is constant, the amount of uranium to lead comparison is just like reading the hands on a clock. The main difference is that you are measuring in calendar years. Hundreds, thousands, and millions of years ago can be figured out because you know what part of the substance is still uranium and what part has changed to lead. Fossils are *DATED* by knowing the age of the rock formation in which they are found.

The Saber-Toothed Tiger

The extinct saber-toothed tigers show up in the fossil record about 40,000,000 years ago. The fossil remains are found in a rock formation where the rocks contain radioactive minerals and are able to be dated. The saber-toothed tigers *became extinct only about* 20,000 to 30,000 years ago.

"Ancient Wing"

One of the most famous fossil finds ever was that of a short-winged creature. The wings had feathers. This fossil also had a lizard-like tail and a beak with teeth. The teeth were like those of a reptile. The evidence shows an organism from 140 million years ago that had features of both reptiles and birds. This organism, called *Archaeopteryx*, got its name from the

Greek for "ancient wing." It was a wobbly flier because of its short wings. It had claws at the front of both wings, probably for catching hold as it made clumsy landings in trees.

Fossil Animal Connections

What is the relationship of animals of the past to the animals of the present? There are connections that can be made. The fossil record shows forms of life in the past that are similar to those living today. On that basis, scientists are able to look at fossil bones and teeth and reconstruct the rest of the animal body. They base information about fossil forms on what they know about the animals of today. Scientists use *DIRECT OBSERVATIONS*. From actual measurements and direct observations, the scientists are able to put two and two together to make *INDIRECT OBSERVATIONS* about the fossils.

Fish fossils have been found in sedimentary rock layers from 530 million years ago. There are records of fish fossils in many different layers of rock. Their history stretched down through the years to the present living species. Most of the record is of impressions of bones in the rock itself. Reconstruction of the whole animal from the information from the bones is based on what is known about fish that are living today and thought to be relatives of those early fish. In addition to the *kind* of fish, the fossil record tells something of the *number* of kinds of fish. The record indicates that there is a greater variety of fishes today than at any time in the lengthy fossil record.

Some other "living fossils" are the *HORSESHOE CRAB*, *OPOSSUM*, and the *COELACANTH FISH*. They are alive today, and also show up in the fossil record. Horses and spiders have also been around in some related form for millions and millions of years.

Coelacanth fish

To show some of the relationships of life forms from the past to animals of the present, the following description of a "model tree" is used. The main trunk is anchored firmly in the ground of the past and shows the beginning of the fossil record. The trunk is wide because it is old (4.5 billion years), but it shows only a small number of different kinds of organisms. The fossil record is slim. However, the fossils that are there represent the oldest record of life on earth. The first fossils do not show the *first* life on earth. They *do* show the first life *with hard parts* like shells and bones. It is certain that simpler forms existed before the first fossil records.

As you go up the trunk to the first branches in this imaginary tree, there is an increase in the variety of life forms. Clams, corals, sponges, and a large number of other invertebrate organisms have left parts of their remains in the rock record.

The most ancient fossil record is speculated to be 3 billion years old. There are skimpy remains of sponge skeletons, some worm holes, and some shells of marine microscopic organisms. Paleontologists continue to work on the problem of what appeared in the fossil record and when. The work is not complete, but the record is clearer every year. Relationships between fossil organisms and living things on earth today continue to be uncovered.

Hunting Fossils Today

Recent expeditions in Thailand have traced the development of vertebrate animals. The fossils have been dated between 240 million years and 65 million years. This is commonly known as the *AGE OF REPTILES*.

Fossils found in the sedimentary layers show that when upheavals formed mountains, the sea was forced to retreat. Then a fresh water lake was formed. The most abundant animal fossils were those of fishes. Scales of old species were identified in the rock history, and toothplates from lungfish. These toothplates, which are formed from fused teeth that are used to crush food, told of air-breathing fish. There are living lungfish in Africa and Australia. Remains of heavily armored reptiles were recognized from their bones and teeth. These reptiles are related to crocodiles, which appear later in the fossil record.

Lungfish

Much of the history of the past is seen in the animals that are on the earth today. It has been estimated that many more species have lived over the long history of earth than there are species alive on earth today.

CHECK YOUR UNDERSTANDING

1. What is the meaning of the term "fossil"?

2. In what kind of rock are most fossils found?

3. What is the origin of fossil fuels?

4. What are two kinds of fossils other than the ones found in sedimentary rock?

5. What does the fossil record show about the teeth and hooves of horses?

6. What is meant by the term "age of fishes" or "age of reptiles"?

7. What is the main idea behind radioactive dating?

8. What is the method for telling the age of the now extinct saber-toothed tiger?

9. What is meant by the term "living fossils"?

10. How old is the oldest fossil record?

Key Idea #2:

- The theory of evolution and natural selection offers explanations for changes in natural populations of living things.

EVOLUTION IS A PROCESS OF CHANGE

Organisms live in populations. All of the members of a population are different. Any *POPULATION* is made up of individuals of the same species. Species are made up of individuals of the same kind and who are able to interbreed and reproduce. *EVOLUTION* is a process of change that occurs through the individuals in a population.

Populations of organisms continue to exist over long periods of time — much longer than any of the individuals in the population. The process of evolution takes place over generations. Each new generation is the product of the genes and chromosomes of the current generation. The theory of evolution describes the results of changes in organisms over long periods that have led to the formation of new species. The theory also offers explanations for the relationship of living things to each other and to fossil remains over long periods of time — millions of years.

Populations, Genes, and Earth Changes

Evolution is a process that involves all of the genes in an entire population of living things. The genes are the determiners of traits and characteristics of the individual organisms in the population. If the genes are changed or altered, that affects the traits and characteristics. Radiation, ultraviolet light, and chemicals in the environment act on genes and cause them to influence changes in the organism. If the changes are useful to the organism, it will cause improved chances for survival. The traits for improved survival are passed on to the next generation by way of genes and reproduction. The usefulness of traits is connected to the kind of environment in which the organisms live.

If the environment changes suddenly, and the conditions for living are not able to be met by a population of organisms, they will die out. The fossil record indicates that such changes have taken place. For example, the dinosaurs were once a dominant species, and now they are extinct. The population of dinosaurs had a pool of genes that influenced the generation of traits that allowed them to survive for millions of years. Whatever the reason for their extinction, they once survived for a long period of time by changing to forms that were able to survive in changing environments.

Splitting a Population

An interbreeding group of animals may be separated into distinct groups by some major change in the earth. A large river or mountain barrier may form so that one group is on each side. The species is now forced to live apart. They do so for millions of years. The group on each side is now made up of a separate gene pool. Generation after generation, there are changes in the genes. Also changes occur in the ways that the genes are grouped and regrouped through sexual reproduction. The changes have to do with different ways of getting food or changes in mating behaviors brought on by chemical changes. Eventually the separate populations become so different that if the occasion arose where the two groups came together, they would no longer be able to interbreed. New species have been formed.

Another example of splitting a population takes place where there is an island off shore from a mainland population. Animals from the mainland travel or are transported across the water. When they live on the island, interbreed, and reproduce for generation after generation, changes occur. Competition for food and for space takes place. Natural changes in the genes take place in the population. The original population becomes more and more different. New and different populations arise on the island. *DIFFERENT SPECIES* result.

Organisms Share Common Traits

The genetic material of all organisms is made up of two chemical substances called DNA and RNA. DNA and RNA are called *NUCLEIC ACIDS*. The nucleic acids are all made of the same basic chemicals. This strong sameness of chemical makeup in all living things indicates a relationship. Creatures of the past show major similarities with animals of the present.

Recent studies of changes that occur in genes of organisms have shown that there is a *STEADY RATE* of change in DNA. This information makes it possible to make estimates about changes in animals and plants based on very precise dating of fossils. Protein changes that take place because of mutations do so at very regular rates. Protein changes are related to DNA changes. These recent studies indicate that the rate of DNA change is as *CLOCK-LIKE* as the process of radioactive decay used to date fossils in sedimentary rock formations.

More and more evidence is becoming available to show the relationships between organisms that are alive today and those that appear in the fossil record. *That these relationships do exist is a big part of the theory of evolution.*

Organisms in general make use of *similar enzymes*. They make use of *chemical energy* in similar ways, and they go through patterns with *similar stages* in their early growth and development. Developing embryos go through stages that are a history of the evolution of the species. Mammal embryos have a stage where the embryo is very fishlike. The basic structures of animals found in the fossil record are similar to structures of modern day organisms.

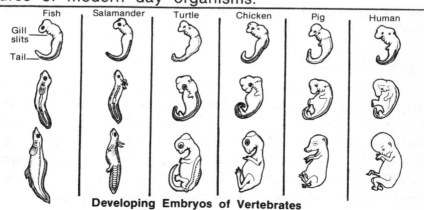

Developing Embryos of Vertebrates

Darwin and Fossils

Charles Darwin was a student of many subjects. He studied *GEOLOGY* (science of the earth) and checked series of fossils that were from *ROCK FORMATIONS* that were known to be very old. He compared them with fossils from younger rock formations. Darwin found similar organisms in both sets of rock formations and noted that similar organisms had *UNDERGONE CHANGE.* The question being asked at the time was *"Do species ever change?"* The fossil evidence showed that species *did* change, and that indeed, *NEW SPECIES* could be formed.

Charles Darwin

Darwin offered explanations for how the changes might have taken place in *POPULATIONS* of organisms. He made detailed observations and studies of changing rock formations in his native country. He made *ESTIMATES* of how long it might have taken for a well-known land formation to have come about. There were different fossils of similar organisms at different layers, and they were much older in some layers than in others.

When all of Darwin's many observations were put together, he had described a *reasonable and logical* set of conditions about the way that many natural events took place. This is what led him to write and publish his famous book, *On the Origin of Species,* in 1859.

Darwin and Natural Selection

A key idea in Darwin's work was that the conditions in *NATURE*, or natural situations, have an influence on which organisms in a *population will survive and reproduce.* This process of natural selection is what causes evolution of a species to happen. *NATURAL SELECTION* takes place as a population produces more offspring than will survive. In populations with very large numbers, there are greater chances of a variety of traits. This greater variation *within the population* results in increased numbers of individuals better suited to survive. These better adapted individuals are the ones that reproduce. The *offspring are also better adapted* to survive, and over long periods of time, there are chances for changes that lead to new *KINDS* of organisms. Each generation has a genetic makeup that equips it to meet the pressures and challenges of conditions in the environment. As the environment changes, the organisms are affected. Those organisms that are best suited with traits that meet the changes in successful ways will survive and reproduce the next generation.

Natural Selection

Young snakes have a tooth located in their upper jaw that is used to cut their way out of the egg shell. This *temporary tooth* is the only means of getting out of the shell. Snakes that are able to do this with ease are better prepared. The selection of young snakes within a population that are successful at breaking out of the egg helps promote the continuation of the species. Unsuccessful young snakes obviously do not live to breed, and their genes are lost from the population. The presence and use of the egg tooth is an example of natural selection acting in a population.

Young birds still in the egg also develop an "egg tooth" to help break the eggshell. The behavior is *innate*. The structure of the egg tooth and its appearance at the time near hatching is controlled by heredity.

Natural Selection at Work

This is a true story about natural selection at work. In Great Britain before the industrial factories caused blackening of tree trunks, there were normal and regular populations of *Biston betularia*, the *PEPPERED MOTH*. Observations of early collections of these insects showed *some light* and *some dark* moths, but mostly lighter moths. In the years after factories were in operation, with their smokestacks spilling out carbon soot, the tree trunks had become dark. Collections of the peppered moth showed a great increase in the number and percentage of dark-colored moths. If you were guessing about why this happened, what would you say?

If you guessed that *natural selection* was at work, you are right! Moths are excellent food for birds. When the tree trunks were lighter, before the factories, any dark moths were picked off by the birds because they were easier to see than the lighter colored peppered moths. As the trees got darker, the situation changed! Now the light-colored peppered moths showed up better on the dark background of the tree trunks. Birds spotted them more easily. Then there were fewer to be collected by the English naturalists.

The environment favored the darker moths, and they reproduced in larger numbers. They became the dominant type of moth in that area. The darker colors of the moth against the darker background kept birds from picking them off as easily as the lighter moths.

A Population of Mice

A story similar to the moth story can be told about a population of white and dark mice living on a white sandy beach. The white mice were a larger part of the total population. They survived at higher rates. They blended into their background and were less often captured for food.

A change in the environment took place. A black flow of lava poured down on parts of the beach. The invasion of the lava favored the darker colored mice in the beach population, and soon there were more dark-colored mice. The white mice were now more visible on the dark lava flow. They were now picked off more often than when the white beach favored them with a better environment for survival.

The color of mice is controlled by genes and heredity. With more dark-colored mice surviving over each generation, what can you say about the future generations in that environment?

If you said that there would continue to be more dark-colored mice, you are right! This change in the nature of the mice population is explained by the theory of NATURAL SELECTION. The theory of the origin of the species by natural selection proposed by Charles Darwin makes use of this kind of example to explain how it works. Other workers before him had made similar suggestions, but Darwin was the first person to do the extensive travel and hard study that was needed to formulate the theory.

Darwin's Finches

When Darwin visited the Galapagos Islands off the coast of Ecuador in South America, he was struck with the variety and distribution of living things. On this set of islands, named for the Spanish word for *tortoise*, he found a variety of birds called finches. There were also finches on the mainland of South America. He observed *fourteen different species of finches* on the islands. Darwin noted that they resembled the species on the mainland. They were different, however. There were differences in the size and shape of the bills. These birds are still known as *"DARWIN'S FINCHES."*

Darwin recorded the fact that the shape of the bill and the size of the bill was different in the fourteen island species. The differences were related to *differences in the diet* of the birds. The birds also stayed in different parts of their island environment — some on the ground — others in the trees.

Darwin reasoned that the island species were probably descended from the mainland species. Separated by the water barrier for a long period of time, numbers of the finch population had undergone genetic changes that influenced the bill. Changes resulted in a bill structure that favored a part of the population, but only in a special part of the island environment.

Success at "food getting" is a major survival trait. As each kind of finch developed and changed, it was better suited to eating a different kind of food item. One kind of finch favored eating small insects, while another finch lived on small cactus seeds. Another finch fed on large seeds and nuts, which it crushed in its heavier and larger beak. A closely related finch fed totally on buds and fruits. One finch species fed on the larvae of insects.

A highly special adaptation to feeding is carried out by one of Darwin's finches that uses a cactus spine to rout out insects that live under the bark of trees. This "tool-using finch" reminds us of the common wood-pecker that uses its highly special beak and tongue to reach under the bark of trees for its insect food.

CHECK YOUR UNDERSTANDING

1. How are populations of organisms a part of the process of evolution?

2. What part did the dinosaurs play in the process of evolution?

3. What happens when a population of organisms is split into two separate groups?

4. What role do islands play in the evolution process?

5. What can you say about the relationship between organisms of today and those of the past?

6. What are some ways that organisms in general are alike?

7. What did Charles Darwin's study of geology have to do with his ideas about the theory of evolution?

8. What is the title of the book that Darwin wrote and published in 1859?

9. How does "natural selection" work? (Use the egg tooth of snakes and birds to explain.)

10. What color of the peppered moth was favored in the area of the soot-covered tree trunks?

Key Idea #3:

- Since its early beginnings in the fossil record, the human species has become the most widespread and the most numerous organism on the planet Earth.

FOSSIL HUMANS

Homo sapiens is the only primate alive today that walks upright on two legs. The earliest fossil record of an animal that walked upright is 3.8 million years old. This fossil, named Lucy, was found in 1974. It shows a skull with an opening for the spinal cord at the bottom of the skull. This is an indication of walking upright. It also shows a pelvis that supports the concept of walking upright on two legs.

In the 1950's, Mary and Louis Leakey, famous fossil hunters, discovered fossil humans that they named *Homo habilis*, not *sapiens*. The brain size was estimated at just more than half the capacity of *Homo sapiens*, 700 cubic centimeters. The Leakeys found simple tools at the site and estimated the age to be about 1.5 million years.

The most recent fossil finds of humans include two groups. The Neanderthal Man has been dated in the last 100,000 years. It has a brain capacity near that of *Homo sapiens* at 1300 cubic centimeters. The other fossil find is that of Cro-Magnon. Cro-Magnons existed about 50,000 years ago along the coast of France. They had fine weapons. They made beautiful cave drawings of the animals that they hunted. Cro-Magnons probably replaced the Neanderthals.

Cro-Magnons are described as modern humans by their genus and species: *Homo sapiens.* They share in the characteristics of modern humans: 1) the largest brains of all primates, 2) walking erect, 3) opposable thumb, 4) tool makers, and 5) use of art symbols.

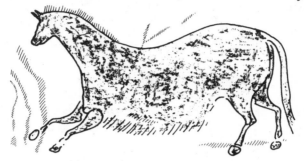

Art found in the cave of Cro-Magnons

Homo Sapiens Today

Modern day humans have a complex language that uses symbols for words. Humans are able to put ideas together and to make written records. Humans have a *CULTURE* that is based on words that are written and spoken. The culture is information that is for human use, about humans, and about everything that has been gained and accumulated as knowledge.

Modern humans have the greatest brain size and capacity of all the primates. With their upright posture, their opposable thumb, their large brain, and with their ability to store information, humans are able to make tools for all purposes. Tool-making is the highest order skill. No other living thing has such a feature.

Humans and the Planet Earth

The human species is the only species that has the brain capacity to think about the earth. Because that is true, humans are, in a sense, in charge of the planet. They know that it is a closed system, and that resources in the earth itself and its waters are limited in supply.

Having studied the science of biology, you have a better knowledge of the many relationships that exist among the living and the non-living materials of the earth. You have a better understanding of the fight to stay healthy, and of the way that plants, animals, protists, and fungi are organized. Your study of ecology and all of the cycles that are involved, makes you think about the earth in a different and more meaningful way.

Although you cannot know everything about everything, you can be a person who solves problems and makes contributions to discussions and debates. You should be able to enter into discussions about genetic engineering, the use of alcohol, drugs, and tobacco, and about the behavior of humans and other organisms. Food webs and food chains are news items in the daily press. You know that oil spills affect organisms, and that lakes, rivers, and bays can become polluted. You know that you can help do something about all of these things. You can act as an individual as well as a member of an action group.

CHECK YOUR UNDERSTANDING

1. What would cause the Leakeys to call their fossil find *Homo habilis* and not *Homo sapiens*?

2. What would be a good reason for not inventing a new naming system for animals and plants from the fossil record?

3. What is the relationship between Neanderthals and Cro-Magnons?

4. What is one of the biggest differences between humans and all other organisms?

5. What can you say about the tool-making ability of humans?

SUMMARY OF CHAPTER 14

Changes that have taken place in organisms over long periods of time come to life as you "read" the fossil record. The record in the rocks, ice, and minerals shows that living things have been around a very, very long time. The record also shows that even though millions of years have gone by, there is a connection between living things in general. The forms and structure of today's animals bear striking resemblances to animals of the near and distant past.

Evolution is a process that involves change. The changes in animals and plants that gave rise to new species is described in the fossil evidence. Layers of sedimentary rock that are extremely old have fewer kinds of organisms. As the layers get younger and younger, there is evidence of a great spread in the kinds and in the numbers of organisms. The great variety of life on earth today is the product of much change in form and structure. The genes of populations have been passed on for generation after generation.

There are many similarities among all kinds of living organisms. The DNA or genetic material itself is made up of the same kinds of chemicals in all creatures. Changes in DNA and in the proteins that they govern, as well as the built-in clocks of radioactive material, give scientists accurate ways of measuring and studying changes that are taking place and that have taken place in the past.

The theory of evolution offers some explanations for how things have changed. A main idea in the theory is that a process of natural selection is at work. Species produce large numbers of offspring and provide a chance for genetic variation to take place. The explanations hold to the idea that species are changeable.

The changes are brought on by the variations of genetic material due to environmental factors such as radiation. Changes also are built into the population gene pool as the genes themselves cross over, break off, and otherwise undergo internal changes.

Homo sapiens has been the most successful of all organisms. The human brain, opposable thumb, upright posture, and stereo vision have been factors in the success of the "thinking animal." The use of language, the development of culture, and excellent tool-making abilities have put humans in front of the line.

Humans have made more changes in the face of the earth and have done more to improve their standard of living than any other living thing. Humans have also done much to disturb the balance of life on the planet Earth. Humans have used the earth's resources with a lack of care and concern. No other organism is so much in charge of its own future and destiny. The problem solving, reasoning, and caring organism known as *Homo sapiens* is well equipped to meet the future with much hope and anticipation.

WORDS TO USE

1. fossils
2. paleontologists
3. anthropologists
4. fossil fuels
5. amber
6. mineralized fossil
7. petrified wood
8. dung
9. adaptations
10. geological timetable
11. radioactive
12. half-life
13. Age of Reptiles
14. Age of Mammals
15. evolution
16. DNA
17. RNA
18. nucleic acids
19. geology
20. natural selection
21. egg tooth
22. Darwin's finches
23. Galapagos Islands
24. *Homo habilis*
25. Cro-Magnon
26. Neanderthal

• REVIEW QUESTIONS FOR CHAPTER 14 •

1. What is the original meaning of the term *fossil*?
2. Where in the world have fossils been found?
3. What is a mineralized fossil?
4. What is petrified wood?
5. How is fossil evidence used to support ideas in the theory of evolution?
6. What can you say about the variety of life forms in the fossil record from 500 million years ago?
7. Some important times in the fossil record are 500 million years ago, 350 million years ago, and 200 million years ago. What does the geological timetable show as the kind of animals that were present in the distant past?
8. How does a radioactive material like uranium help give information about the age of fossils?
9. When did the saber-toothed tigers show up in the fossil record?
10. What do the horseshoe crab, the opossum, and the coelacanth fish have in common?
11. What does the fossil record tell about the numbers of species that have lived on the earth?
12. What is the connection between populations of organisms and the main ideas of evolution?
13. What is evolution about?
14. What happens on an island that helps in the formation of new species?
15. What happens to DNA over long periods of time? What makes that situation like a clock?
16. What are some of the things that Charles Darwin did to arrive at his ideas on the theory of evolution?

• REVIEW QUESTIONS FOR CHAPTER 14 •

17. What is an egg tooth? How does the egg tooth fit with the idea of natural selection?

18. What is the connection between the story of the peppered moth and the idea of natural selection?

19. What was the relationship between species on the mainland of South America and the species on the Galapagos Islands?

20. What are some of the reasons that you can give for the different species of Darwin's finches?

21. What are some of the traits that separate humans from all of the other organisms in the world?

22. What makes tool making such a special characteristic?

INDEX

Learned behavior, 247, 254-258
Leeches, 37
Left atrium, 133
Left ventricle, 133
Lens, 146
Lichens, 73
Life activities, 9
Ligaments, 151
Light energy, 240
Linnaeus, Carolus, 44
Liver, 89, 129
Living things, 17-19
Logical patterns, 257
Lung circulation, 93
Lungs, 91, 136

Mammals, 27, 30-31, 33, 172
Mammary glands, 176
Marrow, bone, 151
Marsupial, 175
Medulla, 143
Meiosis, 167, 206
Membrane, 168
Mendel, Gregor, 199-204, 207
Menstruation, 178
Microbe, 188
Microorganisms, 20-21, 188
Microscope, 20-21, 109
Middle ear, 147
Mildews, 75
Milk, 27, 31
Millipedes, 38, 40
Mineral, 75, 107, 192, 194
Mineralized fossil, 266
Mitosis, 66, 165, 205
Molds, 57, 71
Molecules, 110
Mollusks, 37
Monocots, 46-47
Morgan, Thomas, 209
Mosquitoes, 39, 62
Mosses, 44-45, 52-53
Motor neurons, 99, 144
Mouth, 127-128, 136
Mouthparts, 84-85
Movement, 10
Muscles, 89, 97, 99, 152-153
Mushrooms, 70, 73
Mutation, 212-214, 217

Mutualism, 76

Names, scientific, 29-34
Natural selection, 277-281
Neanderthal man, 282
Nephrons, 138-139
Nerve net, 96
Nervous system, 95, 98, 141-149
Nest-making, 248
Neurons, 144
Nitrates, 238
Nitrifier, 238
Nitrogen, 68, 79, 111, 238-239
Nose, 136
Non-living things, 17-19
Nourishment, 192-194
Nucleic acids, 274-275
Nucleus, 57, 165
Nutrients, 8, 75, 192
Nutrition, 192-194

Offspring, 199
Ooze, 60
Optic nerve, 146
Organisms, 16, 225
Organs, 5
Origin of Species, The 276
Ossification, 151
Outer ear, 147
Ovary, 118, 166
Ovulation, 177
Oxygen, 58, 65, 91, 94, 109-116, 137, 237
Oxygen-carbon dioxide cycle, 116

Paleontologists, 266
Pancreas, 89, 129
Paramecium, 21, 58, 61, 63-66
Parasites, 22, 36, 61, 66, 72
Parental care, 181
Pasteur, Louis, 191
Pathogen, 188
Pavlov, Ivan, 255
Penicillin, 72
Penis, 176-177
Peppered moth, 278
Peripheral nervous system, 141
Perspiration, 139
Petiole, 109